# INDUSTRIAL TRANSITION

# The Dynamics of Economic Space

Series Editor: Neil Reid, University of Toledo, USA

The IGU Commission on 'The Dynamics of Economic Space' aims to play a leading international role in the development, promulgation and dissemination of new ideas in economic geography. It has as its goal the development of a strong analytical perspective on the processes, problems and policies associated with the dynamics of local and regional economies as they are incorporated into the globalizing world economy. In recognition of the increasing complexity of the world economy, the Commission's interests include: industrial production; business, professional and financial services, and the broader service economy including e-business; corporations, corporate power, enterprise and entrepreneurship; the changing world of work and intensifying economic interconnectedness.

# Industrial Transition
## New Global-Local Patterns of Production, Work, and Innovation

*Edited by*

MARTINA FROMHOLD-EISEBITH
*RWTH Aachen University, Germany*

*and*

MARTINA FUCHS
*University of Cologne, Germany*

Routledge
Taylor & Francis Group

LONDON AND NEW YORK

First published 2012 by Ashgate Publishing

Published 2016 by Routledge
2 Park Square, Milton Park, Abingdon, Oxfordshire OX14 4RN
711 Third Avenue, New York, NY 10017, USA

First issued in paperback 2016

*Routledge is an imprint of the Taylor & Francis Group, an informa business*

**British Library Cataloguing in Publication Data**
Industrial Transition: New Global-Local Patterns of Production, Work, and Innovation.
– (The Dynamics of Economic Space)
1. Industrial organization. 2. International division of labor. 3. Economic geography.
4. Space in economics. I. Series II. Fromhold-Eisebith, Martina. III. Fuchs, Martina.
338.6–dc23

**Library of Congress Cataloging-in-Publication Data**
Fromhold-Eisebith, Martina.
Industrial Transition: New Global-Local Patterns of Production, Work, and Innovation/
by Martina Fromhold-Eisebith and Martina Fuchs.
p. cm. – (The Dynamics of Economic Space)
Includes bibliographical references and index.
1. Industrial organization – Case studies. 2. Technological innovations – Case studies.
I. Fuchs, Martina. II. Title.
HD31.F766 2012
338.6–dc23                                                        2012003878

ISBN 13: 978-1-138-25540-1 (pbk)
ISBN 13: 978-1-4094-3121-3 (hbk)

# Contents

# List of Figures

# List of Tables

# List of Contributors

**Oedzge Atzema**, Department of Human Geography and Planning, University of Utrecht, the Netherlands

**Harald Bathelt**, Department of Political Science and Department of Geography & Program in Planning, University of Toronto, Canada, and Zijiang Visiting Professor, Department of Urban and Regional Economy, East China Normal University, China

**Kean Birch**, Department of Social Science, York University, Canada

**Shiuh-Shen Chien**, Department of Geography, National Taiwan University, Taiwan

**Andrew Cumbers**, Department of Geographical and Earth Sciences, University of Glasgow, United Kingdom

**Martina Fromhold-Eisebith,** Department of Geography, RWTH Aachen University, Germany

**Martina Fuchs**, Department of Economic and Social Sciences, University of Cologne, Germany

**Christoph K. Hahn**, Department of Geography, University of Saarland, Germany

**Sebastian Henn**, Department of Geography, University of Halle-Wittenberg, Germany, and Department of Geography & Program in Planning, University of Toronto, Canada

**Yu Ho**, Department of Geography, National Taiwan University, Taiwan

**Wouter Jacobs**, Department of Human Geography and Planning, University of Utrecht, the Netherlands

**Hanno Kempermann**, Department of Economic and Social Sciences, University of Cologne, Germany

**Mingbo Ma**, Department of Urban and Economic Geography, Peking University, Peoples Republic of China

**Dorit Meyer,** Department of Economic and Social Sciences, University of Cologne, Germany

**Andrew K. Munro,** Institute for the History and Philosophy of Science and Technology, University of Toronto, Canada

**Daniel Schiller**, Institute of Economic and Cultural Geography, Leibniz University of Hanover, Germany

**Michael Taylor**, School of Geography, Earth and Environmental Sciences, University of Birmingham, United Kingdom

**Leo van Grunsven**, Department of Human Geography and Planning, University of Utrecht, the Netherlands

**Ton van Rietbergen**, Department of Human Geography and Planning, University of Utrecht, the Netherlands

**Christoph Scheuplein**, Department of Geography, University of Münster, Germany

**Steffen Wetzstein**, School of Earth and Environment, University of Western Australia, Australia

**Jici Wang**, Department of Urban and Economic Geography, Peking University, Peoples Republic of China

# Preface

The world economy and society have rarely before experienced a period of more volatile, erratic and insecure developments than during the recent decade. Industrial production mandates shift to and between lower-cost countries and are relocated to mature economies at an unprecedented pace. Patterns of the international division of labour have become 'fuzzy' and flexible, defying predictions of mid-term trends. Various kinds of socio-economic changes affect different spaces and territories in diverse ways, varying intensities, and with multifaceted outcomes. Scientific scholars who work on global-local development processes from the viewpoint of economic geography, social sciences or economics cannot close their eyes to these major shifts. We need to become aware of the new qualities of fluctuations, their drivers and implications; we must reflect on what they mean for various aspects of regional development. And we need to find new conceptual tools that help to adequately address and analyse relevant dynamics and their manifestations at different spatial scales.

This outlines the motivation which has brought us to compile this book, feeding into the central conceptual framework of 'industrial transition'. We are thankful that a range of colleagues, all of them economic geographers, have been willing to share our concerns and commitment towards putting the emerging challenges into terms. They joined us in the first step, participating in a scientific workshop on 'Industrial Transition – New Patterns of Production, Work, and Innovativeness in Global-Local Spaces' in May 2010 in Cologne, conducted under the auspices of the IGU commission on the Dynamics of Economic Spaces. In this context, cordial words of thanks go in particular to Michael Taylor (University of Birmingham, UK) and Neil Reid (University of Toledo, USA) as chairs of this commission. We appreciate their constant encouragement and support, which guided us from the idea of arranging a scientific event to the completion of this volume, including language editing services by Neil Reid for which we are most grateful. Almost all book chapters draw on this inspiring workshop that brought together scientific scholars from seven countries and four continents. Additional chapters were solicited in order to provide a more comprehensive picture of major facets of industrial transition.

We are indebted to all contributors for reacting to our requests so promptly and for their admirable discipline which substantially supported the progress of this publication endeavour. We would also like to thank the international reviewers for their useful comments on the chapters. Their constructive criticism has helped to further shape our lines of reasoning according to the conceptual objectives.

The shape of what industrial transition actually means may still just vaguely become visible throughout this book. Yet, a range of skilled 'sculptors' of conceptual reasoning in economic geography have started to 'carve out' what, eventually, could turn out to be a fairly imposing monument.

November 2011

MARTINA FROMHOLD-EISEBITH                                    MARTINA FUCHS
*Aachen, Germany*                                          *Cologne, Germany*

Chapter 1

# Changing Global-Local Dynamics of Economic Development? Coining the New Conceptual Framework of 'Industrial Transition'

Martina Fromhold-Eisebith and Martina Fuchs

## Introduction

From the early beginnings of globalization and the rise of economic competition, local-global development dynamics are undergoing permanent, often unprecedented change: Companies compete for international market regions and new market segments, they seek economically priced resources globally, innovate continuously, re-organize their international subsidiaries, and they struggle for preferably optimal corporate locations worldwide. All these industrial activities imply changes in space and place related constellations throughout the world.

At present, there are significant signs of a further increase in international competition, driven by a sequence of economic crises that add pressure on economic actors virtually everywhere. Moreover, we also face the changing nature of the globalized resp. transnationalized production: There are new forms of the division of labour and competencies between corporate locations in different countries. From the regional view, we notice new forms of organization of localized processes in coordination with events at other locations far away. Thus, regions are increasingly entangled in globally coordinated processes. Consequently, new geographical patterns of production and labour appear in highly and less developed countries, challenging managers, trade unions and workers as well as regional policies. Inside of the production sites, there are unclear new patterns of work organization: We perceive a hybridisation of Fordist, post-Fordist and neo-Fordist features of production. Such division of labour comes across in manufacturing; yet partly also in the internationalized service sector, such as software design and various back office activities. High-end tasks and jobs with low qualification requirements of one and the same production process are separated spatially and divided up into different regions of the world. For managers, outsourcing of production steps and service tasks are opportunities that they check permanently, testing out the best international location. Some tests fail, and the jobs are fetched back.

In general, we face an increasing volatility and temporality of tasks conducted at a location, of patterns of distant collaboration, and of firm locations themselves. Regional actors who play their part in the international competition, struggling to create appealing locations for companies, have to react to such diverse and fuzzy trends. International tendencies towards decentralization, regionalization, city-regionalism and localism are changing administrative and governance structures at the national, regional and local levels too. Sets of actors can answer with resilience, resistance or defenceless-passivity. Together, these inter-related developments are generating new challenges and a demand to build capacity for analysis, strategy and policy-making for local and regional development amongst individuals and institutions in the public and private sectors. Local and regional policy-makers have to shape their responses and build the resilience of local and regional economies (Pike et al. 2010).

Obviously, a theoretical reconceptualization is needed that facilitates the comprehension of recent trends. Proposing the concept of 'industrial transition', we suggest an overarching framework that highlights some common ground of various approaches referring to internationalization, globally integrated places and regional development. This chapter elaborates essential ideas of this new heuristic concept and points out how the other book chapters feed into that reasoning. Our key questions are

- In which ways do recent industrial dynamics deviate from conventional explanatory concepts in a time-space perspective? Hence, why do we really need a new conceptual framework beyond the existing ones?
- Which qualities in detail characterize new global-local geographies of production, forming major aspects of 'industrial transition' dynamics?
- In which respects do the chapters of this book illustrate the novel trends, emphasizing crucial features of 'industrial transition'? How do they contribute to a solidification of conceptual assumptions presented in this first chapter?

## Conceptual Foundations of 'Industrial Transition' – From Roots to New Horizons

In economic geography and regional sciences, there is a broad range of theoretical concepts that deal with the issue of spatial dynamics and industrial change. The spectrum ranges from deterministic neoclassical equilibrium approaches and evolutionary thinking to diverse institutional concepts. Beyond doubt several of the new approaches got a kick-start from the debate about the 'new economic geography' claimed by Krugman, Fujita and Venables (1999), on the one hand, and the institutional and the cultural 'turns', on the other hand (Bathelt 2006: 223). When discussing 'industrial transition' we need to venture beyond that, putting even more emphasis on socio-cultural perspectives and time-spatial dimensions (see Ibert 2010).

Before we elaborate on these issues, however we must concede that we are not the first and only ones to use the 'transition' term from an economic geography perspective. Alluding to ideas of sustainability and the reconfiguration of sectors through long term social and technological shifts, so-called 'socio-technical transitions' have received increasing attention in economic geography, including insights from evolutionary economics, organization studies or science and technology studies (Coenen et al. 2010, van den Bergh et al. 2011). In these works some reference has been made to industry and cluster life cycles and territorial industrial transformations as well. Yet, the focus is set on (disruptive) technology formation and socio-technical transformations, inspiring research on technology's meaning, constitution, impact, change, and diffusion. Research in this field is tuned to questions about environmental and social sustainability, looking at power relations, place-specific contingencies, and a range of practices, relationships, and actors. While both the socio-technical transition and the industrial transition notions consider multi-scalar processes and recent insights from evolutionary, institutional and relational economic geography approaches, our own concept is less focused on technological changes driven by sustainability considerations and their societal implications than its distant relative. Instead, our framework rather emphasizes industrial and production related dynamics across places and scales, which strongly influence the activities of economic units from nations down to the level of individual workforce.

*'Industrial Transition' as a Time-Spatial Concept*

When picking up the term 'industrial transition' to characterize the recent industrial change, we draw on Michalet (2007: 14) who proposes a 'grande rupture' in the global economy, explicitly using the expression 'phénomène de transition'. He refers to the new divisions of labour, competencies and innovative patterns of local-global coordination, including topics of international finance and regulatory aspects of coordination and control.

In our understanding of 'industrial transition', the expression 'transition', rooted in the Latin word 'transire', offers the advantage that the term does not relate to a specific concept of time, as cycles or evolution; the concept is open to a multiple and compound perspective, as this seems appropriate for the described phenomena. With regard to current industrial changes, the term is less theory loaded than 'transformation', 'adjustment', 'alteration', or 'shift'. Yet, in general, we refer to time as historical time associated with complex socio-economic development, not in an abstract and deterministic sense.

The idea of 'industrial transition' is open to different perspectives on time; yet, it is not indifferent. The basic idea is that in our recent historical period of globalization, we find different kinds of dynamics occurring simultaneously: Concurrently, we observe cyclical processes, evolutionary development, change framed by institutions, and alteration or transformation resulting from crisis. Remarkably, recent approaches of economic geography seldom deal with the issue of crisis (Fromhold-Eisebith and Eisebith 2011). The comprehensive *Handbook of Evolutionary Economic Geography* (Boschma and Martin 2010) is to be regarded as an indication, as not a single chapter of the handbook deals with the subject of crises. The reason may be that the evolutionary perspective is directed on the gradual and ongoing self-transformation of the economy from within, not on sudden and deep shocks, or recessions resulting from cyclical processes. Even though the notions of emergence and unforeseeable, 'non-ergodic' developments, 'critical junctures' and 'historical accidents' implicitly include the notion of abrupt and profound shocks (MacKinnon et al. 2009: 131–132), evolutionary economic geography does not deal conceptually with crisis (Martin and Sunley 2010: 65–75). In contrast, regulation school, which has become a bit out of fashion today, deals with the crisis (Aglietta 1976, Boyer 1995), including problems resulting from the absence of institutional regulation in less developed countries, criticizing the 'fordisme incomplet' and 'sous-fordisme', respectively, in the branch-plant economies still prevailing in many countries of the Global South (Lipietz 1986).

In sum, at present we find different kinds of dynamics and change simultaneously shaping the 'industrial transition', which have not adequately been conceptualized from an overarching viewpoint before: Long lasting cycles of booms and recessions, impacted by basic innovations; product life cycles, partly explaining the international division of innovation activities and their effects on labour and competencies; evolutionary processes proving that legacy plays a significant role and emphasizing that innovation is part of changing socio-economic systems; institutional change referring to different levels, various issues, different intensities and directions of change; last not least, crisis leading to adjustment or transformation. As becomes apparent from the contributions to this book, the recent economic downturn can even be regarded as a kind of catalyst for the amalgamation of various types of industrial dynamics. Although 'industrial transition' should not to be seen as an immediate concomitant of the world economy crisis, the latter has, without doubt, further sharpened the profile of changes captured by our new conceptual notion.

The contemporaneity of diverse ways of change comes across in space, hence the concept of space also needs some deliberation. In contrast to other concepts, the idea of 'industrial transition' is open to different perspectives on space, too. Considering the widespread critique that space is not a simple 'container' of actors and objects that have their place in a physical three-dimensional 'box' or that move or migrate between places, new notions have been developed, better fitted to recent theoretical insights as well as to empirical trends of globalization. Like concepts of time, there are no universal concepts of space that suit all matters: Obviously, 'glocalized' relationships are important, and we talk daily about global-local relations. Yet, the widespread dualistic view of 'the global' and 'the local' is unsatisfactory to understand the new patterns on different spatial scales.

Referring to multi-scalar events and processes, heuristic concepts of space appear as layers, as concentric circles or as roots. A current notion of space is the rhizome, or the metaphor of wormholes, referring to passages that cut through space, relating to the interconnectivity of the world economy, the diverse directions of processes, and broadly corresponding to the 'relational' approach (Herod 2003: 238–41).

## Roots of the 'Industrial Transition'

Of course, the concept of 'industrial transition' draws on some antecedents. Major starting points lie in the notions of global production networks, in the regulation approach, and in insights into regional restructuring.

### Global production networks

As a spatial perspective on complex multi-scalar interrelations and regional networks, which comprises organization, actor, interaction, contextuality and global-local knowledge creation (Bathelt 2006: 224–30), the relational approach is in close range to the idea of 'global production networks'. Important links to the recent debate about the global production networks go back to Gereffi (1999), who started a vivid discussion in the last two decades. They were focussing on global commodity chains and their effects on regions in less developed countries. Before, Fröbel, Heinrichs and Kreye (1980) had already brought up the subject of structural unemployment in industrialized and less developed countries, resulting from the – at that time – new international division of labour.

Today, the discussion has shifted from commodity chains and value chains towards more complex global production networks (Hess 2008). Thereby, the focus has moved from unilinear linkages towards manifold interwoven connections. Furthermore, the dual idea of 'core economies' and 'less developed countries' has dissolved, as tasks of the global production networks are carried out in dissimilar countries and regions, which can be better characterized by ideas of the varieties of capitalism (Hall and Soskice 2001) or different institutional settings.

Additionally, control mechanisms have become an important issue in the recent debate too. Besides governmental policies and supra-governmental procedures, as enacted by the European Union, 'global governance' performed by private and public actors as well as by non-governmental (humanitarian or environmental) organizations play an essential role in recent studies (Hess 2008).

### Regulation and modes of production

While 'global governance' refers to the transnational, global level, there is also a broad debate about regulation, deregulation and re-regulation of socio-economic spatial relations on national and sub-national level (Jessop and Sum 2006). Focussing on regulation on the local scale, especially on the shop-floor level, we arrive at an overlapping research field of economic geography, labour geography and labour sociology. With regard to labour organization, there is a broad interdisciplinary discussion about the situation on the shop floor level, also related to the debate about the varieties of capitalism (Herrigel and Wittke 2005). We are facing a multiplicity of modes of production organization.

If we only refer to automobile companies as an example, we find different 'national' modes of production in the factories of the USA, Japan and the European countries, as the rather 'Taylorist' American way of production, the 'lean' Japanese concept of Toyotism, and the different kinds of production organization in the French, German, Italian etc. companies – accompanied by different cross-combinations between the different modes in the various companies and subsidiaries (Boyer and Freyssenet 2000).

The pattern of the international division of labour and competencies is fuzzy: Still, we frequently find repetitive 'Taylorist' work in the core economies. As well – in the core economies and, to a less degree, also in non-core locations – there is a broad range of 'post-Fordist' work organization, including teamwork and other kinds of workers' participation, high qualification requirements and considerable responsibility of the employees. Besides, there are 'neo-Fordist' ways of work organization, combining flexible process organization with standardization and modularization in various countries. And all different kinds of the organization of production resp. work can be combined with varying patterns of in-house production, outsourcing and international offshoring.

Such different modes of production organization are not only a simple result of managerial strategies; in several countries they are an expression of legislation as well as an outcome of the bargaining power between works councils, trade unions and other different interest groups. Also the interrelation of central management and local management of the subsidiary shapes the specific modes of production. Local competencies and workers' qualifications play an important role (Fuchs 2005, 2007, 2008).

*Regional Restructuring*

A further origin of the 'industrial transition' idea is the literature about regional restructuring. There is a broad debate about deindustrialization and old industry regions in Europe and the USA, as well as in less developed countries. And there are many concepts dealing with old buildings and estates, protecting the industrial heritage, giving them new prestige and creating a new 'cultural' branding of regions (Pike 2009). We recognize various agendas of regional adaptation and resilience (Pike et al. 2010), reaching from the restructuring of old coal and mining areas and other former manufacturing sites to diverse kinds of waterfront redevelopment.

Other approaches analyse the new formation and modernization of companies and regions; such research puts a strong focus on start-up and spin-off companies as well as collaborative arrangements between research and development (R&D) and production. The mutual interdependencies of innovation dynamics and regional transformation have been widely acknowledged (Fromhold-Eisebith 2009). This goes in line with realizations of the scale-crossing nature of innovation processes and systems (Fromhold-Eisebith 2007). These debates underscore the new qualities and extended range of various actors' interaction that mark contemporary regional industrial change.

Taken together, the depicted conceptual approaches have already highlighted several important development trends from a time-spatial perspective. We argue, however, that something has been lacking in terms of conceptualization, which can form a crucial complement to what has been discussed before. We advocate an overarching framework which, in the sense of an eclectic heuristic approach, creates some kind of common ground for various concepts on volatile global-local industrial dynamics. More emphasis should be set on the recently emerging qualities of internationally embedded industrial processes, especially with respect to accelerating, erratic, circular and heterogeneous developments and to diverse spatial patterns of activity. We also need to capture more adequately the various aspects of regional development that are – and will be – particularly affected by those trends. Implications for policies, and of policies, need to be considered too, in order to allow the new concept to offer not only some guidelines for development analysis, but also for the implementation of appropriate measures.

**Towards a Conceptualization of 'Industrial Transition'**

Habitually, it is very difficult to distinguish the conceptual features of the most recent development period; *ex post* assessments of 'completed' processes are much less complicated to conduct. In addition, the international division of labour and competencies in manufacturing has grown increasingly volatile recently, driven by a truly complex fabric of actor strategies and multidirectional dynamics, which renders any attempt at conceptualizing this setting fairly speculative. Thus, the heuristic framework of the 'industrial transition' cannot avoid being rather fuzzy and vague at this stage. Nevertheless, we can already point out some issues that are to be conceived as central pillars of our approach. In some respects, they form the foci of the heuristic lens which we use through the 'industrial transition' notion. The twelve conceptual features and aspects elaborated on in the following are also – in different guises and to varying degrees – illustrated by the case study chapters compiled in this book. These chapters are introduced in more detail further below.

A first major attribute of the 'industrial transition' approach is that it acknowledges the *incessant continuity of change* – embodied in the meaning of the term 'transition' – and that it is not possible to draw clear distinctions between certain development phases nowadays. Manufacturing companies and their service partners operate in relentlessly transforming business environments. Consequently, their capabilities to adopt, adapt and flexibly change strategies lie at the heart of their abilities to successfully compete internationally. This is associated with the establishment of new routines of adaptability and adjustment, which forms an almost paradoxical set of challenges. Due to highly insecure conditions of operation, companies increasingly compete through complex blends of tangible and intangible assets, which are meant to enhance corporate resilience.

This means that firms of all sizes also draw on their history and heritage, place-based associations, brands and design. Emphasizing such intangible forms of competitiveness offers new ways to cope with ever changing contexts.

Second, our conceptual framework regards regions as relational settings, going beyond macro-economic analyses of trade and other kinds of linkages. 'Industrial transition' also leaves micro-economic internationalization theories behind, which analyse intra-firm connections and the links between suppliers and clients only as purely economic networks. Our approach, in contrast, considers the development of companies and sectors as integrated into *socio-economic relations between different regions of the world economy*. Thus, 'industrial transition' explicitly looks at connections between locations in core economies, newly industrialized countries and less developed countries, revealing new forms and flexible means of the international and interregional division of tasks and mandates.

Third, by analysing internationally related places, research conducted on 'industrial transition' refers to headquarters, production plants, sites of R&D centres and other competence centres. Accounting for the differing characteristics of these activities, we comprehend that the *regional effects* of operations in such multi-location settings are diverse and dynamic. International shifts often lead to processes of *upgrading;* yet, they still draw on cheap labour in less developed countries. In some cases, linking up to global value chains may even lead to *downgrading* and the exploitation of labour and natural ecosystems. The 'industrial transition' framework recognizes the potential for contradicting dynamics within corporate, national and regional boundaries.

Fourth, our conceptual approach aims at dissolving harsh distinctions between modes of production. This refers in particular to observations of a *hybridisation of Fordist and post-Fordist features of production*. In line with that, we see the emergence of new forms of organization of localized processes, and new forms of coordinating events and activities conducted at distant places. The corresponding shifts along the continuum of highly standardized and highly customized production also play a major role for regional impacts. There is an increasing volatility and temporality of tasks conducted at one location, of patterns of external collaboration, and of firm locations themselves.

A fifth attribute of the 'industrial transition' approach is its sensitivity concerning terminology. We must, for instance, be careful when talking about *'global' relations* with regard to the *topics of control, coordination and interconnections*. We can speak of 'global' relations when considering constellations in a broader sense. More specifically, 'transnational' relationships refer to control and coordination across national borders, frequently affecting the institutional settings of involved nations. While the term 'transnational' implies interconnections and blurred borders, 'international' rather refers to firm-to-firm (or to state-to-state) interactions (Djelic and Quack 2008: 300).

Sixth, we acknowledge that international or transnational relations may appear at *different spatial scales*. They can emerge either as more or less worldwide interdependencies or between certain locations, linking very few, partially neighbouring places. For example, worldwide *offshoring* was already a topic of discussion in the 1980s (Fröbel, Heinrichs and Kreye 1980). Recently, *nearshoring* of West European companies towards Central Europe has also become a key issue. Relating to nearshoring activities in Europe, Baldwin (2006) suggests the paradigm of a 'second unbundling'. Before, international competition was manifested as a contest between companies, sectors and locations. Now competition and rivalry also take place between individual workers performing similar tasks in factories at different European locations, which sometimes belong to the same company.

Seventh, with regard to manufacturing sectors, the 'industrial transition' concept suggests a deeper understanding of the *'new, old economy'* (Ettlinger 2008). 'High tech' and 'low tech' are less regarded as analytical terms than as the wording of industrial policy. In fact, we find high tech and R&D activities in the countries of the economic triad as well as in some less developed countries. Furthermore, low tech is still a considerably widespread (yet frequently denied) mode of production in the mature core economies (Hirsch-Kreinsen 2005). This forms another perspective on the hybridisation of Fordist, post-Fordist and neo-Fordist features of production and services in different regions and on different scales of the globalized economy.

The eighth 'industrial transition' aspect refers to broader systemic processes of *adaptation, resilience and crisis response* on the regional scale, involving entire sets of different actors. Mechanisms of resilience (Pike et al. 2010) have been particularly challenged by the economic crisis, which has not only affected industrial regions in the mature economies but was also transmitted to a broad range of places elsewhere through far reaching production and service supply linkages. Conversely, responses also need to be organized along these channels. The crisis is conceived as a catalyst that renders the volatility of dynamics, captured in the 'industrial transition' notion, particularly visible.

Ninth, the secular trends towards increasing mechanization and electronification in industrial production as well as towards sophisticated tasks challenge many regions with the problem of perpetual *change of qualification requirements*. For regional actors, technical training as an adjustment measure and the threatening problem of jobless growth leading to regional unemployment are important issues. For many urban areas especially, the growing crowd of unqualified workers leads to trouble. Since the separation between 'high end' tasks undertaken in the traditional core economies of the triad nations and 'low end' tasks undertaken in newly emerging economies has become increasingly blurred, we find fragmented labour in different world regions.

With regard to service industries, the tenth focus of 'industrial transition' picks up the discussion about financial services (Michalet 2007). Financial actors, transactions and strategies are essentially perceived as major drivers and sculptors of transition processes in various industrial sectors. Additionally, this view links to studies about *global cities and strategic transnational networks* between the nodal centres of decision making (Sassen 2001). The areas between such nodes are often described as forgotten or lost regions. However, also in the centres resp.

in the global cities we find large numbers of less qualified service jobs, often filled by migrant workers who sustain the new service industries and the high-end jobs.

Eleventh, the 'industrial transition' framework deals with the concept of the region per se. It goes beyond the idea of regional embeddedness and takes new approaches seriously, including manifold, often combined views on institutions, actors and practices, discourses and interpretation, power and dependency. Legacy, as stressed by evolutionary approaches, plays a strong role in the sense of *place dependency* (Martin and Sunley 2010: 63).

In terms of a twelfth attribute, *governance, regulation and control* represent important issues of the 'industrial transition' approach. 'Corporate governance' is a current, yet for many actors unsatisfactory, answer to the profound ongoing changes and challenges. Regional, national and supra-national governments face the escalating power shift towards private corporations. Not only do global manufacturing and some service industries challenge autonomous politics on different scale; additionally, finance and banking play an increasingly important role too.

## Plan of the Book

The chapters of this book have been selected due to their highly illustrative value with respect to major features of 'industrial transition', as outlined above. As the conceptual framework addresses truly 'globalized' trends, the contributions to this book highlight dynamics in different parts of the world and in various fields of industrial activity. Still, all these developments must be considered as being linked in some way, in line with the systemic underpinnings of 'industrial transition' that connects between places, sectors, and several economic and social processes. The following sections underscore the ways in which the authors contribute to the overarching objectives of this book, and which specific facets of 'industrial transition' are revealed.

In Chapter 2, Michael Taylor sets the broader agenda by discussing how economic geographers can tackle the scientific challenges associated with new qualities of interdependencies between industry, economic growth and recession. Exposing 'forgotten issues' in current debates, he arrives at some constructive criticism which provides useful guidelines for further conceptualizing the 'industrial transition' in expedient ways, and for relevant empirical research based on these grounds. He suggests an agenda for action that touches upon various 'sore points' of contemporary institutional and spatial imbalances of socio-economic development that require specific attention. Eight aspects – from new interpretations of manufacturing to people-focused issues – deserve more emphasis in future economic geography research, enabling us to better capture the true nature of 'industrial transition'.

In Chapter 3, Christoph K. Hahn elucidates major sector specific implications of 'industrial transition' in terms of the mobilization of regional knowledge, using the example of the automotive industry in the border-crossing 'Greater Region SaarLorLux'. According to his empirical investigations, the recession has increased pressure to react especially among small and medium sized suppliers in the automotive value chain. Additionally, other drivers of technological change play a strong role in firm restructuring, including government requirements of environmental protection. Under conditions of growing competition, the companies follow diverse organizational strategies, slightly increasing the extent of cross-border interaction. However, overall cross-border cooperation in the 'Greater Region' is still not well developed due to low levels of functional proximity between industrial agents.

The subsequent Chapter 4 by Christoph Scheuplein on industrial restructuring by financial investors discusses a major conduit between transition dynamics in different process fields. He reveals how private equity investment geared towards buy-and-build strategies has recently affected and transformed industries, looking at the German automotive supply industries. It is shown that the amalgamation of various forces, such as the interaction of private equity investments and the crisis, produces distinct patterns of 'industrial transition'. These patterns are shaped by specific world economy constellations, management routines of private equity firms, and the structural conditions of the regarded industrial sector. Obviously, private equity involvement increases the vulnerability of industries in the crisis and must therefore be considered as an influential destabilizing factor driving 'industrial transition'.

As Martina Fuchs and Hanno Kempermann argue in Chapter 5, drawing on the case of the German mechanical engineering industry, the paradigm of 'flexible specialization', once considered a major competitive asset in the 'Second Industrial Divide' and centrepiece of post-Fordist production (Piore and Sabel 1984), has changed considerably under the pressure of global competition. From a conceptual angle and based on qualitative empirical research, the authors show that too much specialization can lead to cost disadvantages and 'lock-in'. In the 'industrial transition' era, the urge for cost-cutting standardization intervenes with strategies of flexible reaction. Firms take advantage of newly emerging options in their institutional environment relating to labour flexibility and equity capitalisation in order to cope with changing market demands and innovation requirements. New forms of (cooperative) flexibility mark the 'industrial transition'.

In Chapter 6 on the organization and representation of temporary workers, Dorit Meyer explores certain labour market related implications of 'industrial transition' and decline, in terms of a dwindling interest of employees in labour union membership. Using the empirically investigated example of a large German metal workers union (IG Metall), she illustrates how these organizations react to recent labour market shifts by reorienting recruitment efforts towards a hitherto neglected, however growing labour segment: temporary workers. The conceptual notion of dynamic capabilities serves to explain how and under which conditions the union transforms its strategies, involving the interplay of various spheres of activity. A 'geography' of union strategy transition becomes visible, shaped by place-specific variations in combining assets from the three hierarchical levels that mark the German union.

Kean Birch and Andrew Cumbers elaborate on changing economic governance and institutional arrangements in the Scottish life sciences in Chapter 7. They start their reasoning by acknowledging the uncertain present and future conditions of operation even in sophisticated sectors. In the light of the financial crisis, regional politicians and planners support knowledge-based or knowledge-driven industries. This also happens in less favoured regions, allocating an important role to Scottish life sciences. The authors argue that a successful 'industrial transition' towards knowledge-based sectors can only be secured through a reconfiguration of economic governance along global value chains. Additionally, the grounding of these chains within specific institutional arrangements at the local and regional level is needed. Accordingly, regional policies have to take account of the international as well as the regional resp. local scales.

Several chapters in this book refer to strategic coupling between firms. Chapter 8 by Leo van Grunsven, Wouter Jacobs, Oedzge Atzema and Ton van Rietbergen explicitly take up this aspect by focusing on the relations between multinational subsidiaries and Knowledge Intensive Business Services (KIBs). The latter are conceived as important integrating agents that, on the one hand, become parts of multinational companies, embedding them in the host region, and, on the other hand, also form crucial elements of regional innovation systems. This illustrates how, in the 'industrial transition', economic actors need to bridge scales and kinds of activity in order to prosper. The authors develop a model of the different ways in which strategic coupling can be used by multinationals. Strategic engagement with KIBs emerges as a useful relational asset for firms and their surrounding regions.

In Chapter 9, Harald Bathelt and Andew K. Munro emphasize change and regional growth dynamics, discussing the choice between intra-firm adjustment and organizational ecology. The authors show how the recession has triggered firms' responses with respect to these two choices. They provide evidence that global financial markets influence manufacturing activities and, consequently, their regions of location. The reasoning draws on a modified organizational-ecology conception that combines intra-firm adjustment with firm formation processes at a regional level. The results suggest that technological change and regional growth can be achieved when technological impulses, as those induced by start-ups, are combined with learning and adaptation processes of incumbent firms.

Starting with Chapter 10, dynamics involving newly industrialized and less developed countries are regarded against the backdrop of 'industrial transition'. Sebastian Henn discusses major interdependencies between a global shift in production and transnational migration, focussing on transnational entrepreneurs of diamond manufacturing. Looking at activities that link certain places in India, Belgium and the USA, he shows the relevance of specific skills in immigrant businesses. The chapter makes clear that not only high tech competencies, but also soft skills and social networks play an important role for connecting between formerly distant regions and for creating new local clusters in the network.

Interpersonal linkages and social networks have gained much attention in recent research on globalization and appear to strongly shape spatial patterns of interaction in the 'industrial transition'.

The role of production shifts to China deserves particular attention within wider processes of 'industrial transition'. Using the example of Western musical instrument production, Chapter 11 by Crison Chien, Jici Wang, Yu He and Mingbo Ma sets a spotlight on these dynamics. Their depiction of the formation and transformation of the Chinese piano industry over the past decades elucidates how even goods formerly regarded as a stronghold of Western cultural orientation and manufacturing focus have now experienced a dramatic location change. These dynamics, which have made China the world's largest producer of pianos, draw on the co-evolution of demand shifts of the growing Chinese middle class towards Western classical music and subsequent shifts of instrument production. Changes relating to society, market and manufacturing interact in producing global-local development dynamics.

Subsequently, in Chapter 12, Daniel Schiller deals with the spatial and organizational transition of the Greater Pearl River Delta, which has turned into an East Asian high-growth region and major production hub of global electronic industries in the last thirty years. The chapter tells that firms from Hong Kong and the Pearl River area are increasingly taking up higher positions in the electronics value chain. Within the region a distinct spatial clustering takes place, with a shift from hierarchical towards market-based relations. At the same time cross-border relations are being transformed due to a growing relocation of domestic and foreign firms within the Pearl River Delta, which further supports clustering tendencies of electronic firms in some cities and districts. The chapter underscores that China is on a path of economic upgrading and becoming a core player of the world economy.

In Chapter 13 Steffen Wetzstein focusses on the nexus between economic competitiveness, urban-centred globalization (termed 'glurbanisation') and 'soft' territorial governance strategies in Australia and New Zealand. For four cities in the wider 'Australasia' region he discusses the importance and influence of discursive practices and imaginaries, which are used by politicians and planners to make their regions more competitive in the world economy. In line with that he highlights the significance of benchmarking and ranking exercises and positions respectively, for the global competition of city regions these days. Accordingly, 'industrial transition' does not only imply the transition of economic relations, but also the collective mental construction of regions as highly ranked investment locations.

In the final chapter, Chapter 14, we pull together the strings laid out throughout this book and derive some overarching insights from the wealth of empirical material. This refers, first and foremost, to the most prominent and visible recent trends and patterns of 'industrial transition' as a contemporary phenomenon. We find promising novel perspectives on the linkages between the macro, micro and meso levels of development as well as on the various economic spheres and sectors involved. Secondly, the book chapters offer a range of inspiring terminological and conceptual ideas on how to further refine the 'industrial transition' framework. And thirdly, corporate networks and industrial locations are faced by mounting risks, which raises the question of how to adequately react at different spatial scales through government and governance. Hence, ideas for future empirical, theoretical and political agency can be derived.


## References

Aglietta, M. 1976. *Régulation et crises du capitalisme. L'expérience des Etats Unis*. Paris: Calmann-Lévy.

Baldwin, R. 2006. *Globalisation: the great unbundling(s)*. Economic Council of Finland. Available at: http://hei.unige.ch/~baldwin/PapersBooks/Unbundling_Baldwin_06-09-20.pdf [accessed: 18 November 2011].

Bathelt, H. 2006. Geographies of production. Growth regimes in spatial perspective 3 – toward a relational view of economic action and policy. *Progress in Human Geography* 30(2), 223–36.

Boschma, R. and Martin, R. (eds.) 2010b. *The Handbook of Evolutionary Economic Geography*. Cheltenham, Northampton: Edward Elgar.

Boyer, R. 1995. Aux origenes de la théorie de la régulation, in *Théorie de la régulation. L' état des savoirs,* edited by R. Boyer and Y. Saillard, Paris: Découverte, 21–30.

Boyer, R. and Freyssenet, M. 2000. *Les modèles productifs*. Paris, Repères: Découverte.

Coenen, L., Benneworth, P. and Truffer, B. 2010. *Towards a spatial perspective on sustainability transitions*. CIRCLE Electronic Working Papers 2010/8, Lund: Lund University.

Djelic, M.L. and Quack, S. 2008. Institutions and Transnationalisation. In: *Handbook of Organisational Institutionalism,* edited by R. Greenwood, C. Oliver, R. Suddaby and K. Sahlin-Andersson. Los Angeles: Sage, 299–323.

Ettlinger, N. 2008. The predicament of firms in the new and old economies: a critical inquiry into traditional binaries in the study of the space-economy. *Progress in Human Geography*, 32(1), 45–69.

Fröbel, F., Heinrichs, J., and Kreye, O. 1980. *The New International Division of Labour: Structural Unemployment in Industrialised Countries and Industrialisation in Developing Countries*. Cambridge: Cambridge University Press.

Fromhold-Eisebith, M. 2007. Bridging Scales in Innovation Policies: How to Link Regional, National and International Innovation Systems. *European Planning Studies* 15(2), 217–33.

Fromhold-Eisebith, M. 2009. Space(s) of Innovation – Regional Knowledge Economies, in *Milieus of Creativity. An Interdisciplinary Approach to Spatiality of Creativity*, edited by J. Funke et al. Heidelberg: Springer, 201–18.

Fromhold-Eisebith, M. and Eisebith, G. 2011. Economic crisis and innovativeness – exploring geographies of impact. Accepted for publication in *Erde,* 142(4).

Fuchs, M. 2005. Internal Networking in the Globalising Firm, in *Linking Industries Across the World,* edited by C. Alvstam and E.W. Schamp, Aldershot, Burlington: Ashgate, 127–46.

Fuchs, M. 2007. Product Upgrading and Survival: The Case of VW Navarra, in *International Business Geography*, edited by E. Wever and P. Pellenbarg, London: Routledge, 216–33.

Fuchs, M. 2008. Subsidiaries of Multinational Companies: Foreign Locations Gaining Competencies? *Geography Compass*, (3)2. Available at: http://www.blackwell-compass.com/subject/geography/article_view?highlight_query=fuchs&type=std&slop=0&fuzzy=0.5&last_results=query%3Dfuchs%26topics%3D%26content_types%3DALL%26submit%3DSearch&parent=void&sortby=relevance&offset=0&article_id=geco_articles_bpl161 [accessed: 18 November 2011].

Gereffi, G. 1999. International trade and industrial upgrading in the apparel commodity chain. *Journal of International Economics* 48(1), 37–70.

Hall, P.A. and Soskice, D. 2001: Varieties of Capitalism: The Institutional Foundations of Comparative Advantage. Oxford: Oxford University Press.

Herrigel, G. and Wittke, V. 2005. Varieties of Vertical Disintegration: The Global Trend Toward Heterogeneous Supply Relations and the Reproduction of Difference in US and German Manufacturing, in *Changing Capitalisms: Internationalisation, Institutional Change and Systems of Economic Organization,* edited by G. Morgan, E. Moen, and R. Whitley. Oxford: Oxford University Press, 312–51.

Herod, A. 2003. Scale: The Local and the Global, in *Key Concepts in Geography,* edited by S. Holloway. London: Sage, 229–47.

Hess, M. 2008. Governance, Value Chains and Networks. *Economy and Society,* (37)3, 452–59.

Hirsch-Kreinsen, H. 2008. Low-Tech Industries, in *Low-Tech Innovation in the Knowledge Economy,* edited by H. Hirsch-Kreinsen, D. Jacobsen and S. Laestadius. Frankfurt: Peter Lang, 147–66.

Ibert, O. 2010. Relational distance: Socio-cultural and time-spatial tensions in innovation practices. *Environment and Planning A*, 42(1), 187–204

Jessop, B. and Sum, N.L. 2006. *Beyond the Regulation Approach. Putting Capitalist Economies in their Place.* Cheltenham: Edward Elgar.

Krugman, P., Fujita, M., Venables, A. 1999. The Spatial Economy – Cities, Regions and International Trade. Cambridge: MIT Press.

Lipietz, A. 1986. *Mirages et miracles. Problèmes de l'industrialisation dans le tiers monde.* Paris: Découverte.

Martin, R., Sunley, P. 2010. The place of path dependence in an evolutionary perspective on the economic landscape, in *Handbook of Evolutionary Economic Geography,* edited by R. Boschma and R. Martin. Cheltenham, Northampton: Edward Elgar, 62–92.

Michalet, C.-A. 2007. *Mondialisation, la grande rupture*. Paris: Découverte.

Pike, A. 2009. Geographies of brands and branding. *Progress in Human Geography*, 33(5), 619–45.

Pike, A., Dawley, S. and Tomaney, J. 2010. Resilience, adaptation and adaptability. *Cambridge Journal of Regions, Economy and Society*, 3, 59–70.

Piore, M.J. and Sabel, C.F. 1984. *The Second Industrial Divide. Possibilities for Prosperity*. New York: Basic Books.

Sassen, S. 2001. *The global city*. New York, London, Tokyo, Princeton: Princeton University Press.

Van den Bergh, J.C.J.M., Truffer, B. and Kallis, G. 2011. Environmental innovation and societal transitions: Introduction and overview. *Environmental Innovation and Societal Transitions*, 1, 1–23.

# Chapter 2

# Industry, Enterprise, Economic Growth and Recession: Forgotten Issues in Economic Geography

Michael Taylor

## Introduction

As technology has changed and the global economy has developed, all national and regional economies are faced with issues of industrial transition. But now, regional economies, worldwide, are facing the brutal realities of contemporary capitalism in recession. Since the 1980s, financialization and the commodification of money, coupled with the growth of the corporate sector and processes of deindustrialization have turned so-called 'developed' western economies into distorted, service-based, consumer-driven entities. Recession, however, has brought the debt-fuelled consumerism of these economies to an end. And with it, the expansion of public sector employment, used to compensate for the loss of jobs in production exported to low cost countries, has also come to an abrupt end. The financial crisis and debt contagion of the Eurozone illustrates these processes only too clearly. At the same time local economies are trying to create jobs and employment, to foster businesses and to be innovative as away of securing growth. There has, as a consequence, never been a more important time for economic geography to contribute to thinking on economic change and industrial transition. The empirical foundation of the discipline has never been more important to counter the monoistic theoretical thinking of free-market economics that has dominated political and economic thinking since the 1980s.

Without change, the future for many economies, especially in Europe, is grim. Public debt is higher than in 1929, but with a financial system driven by speculation that is itself supported by public funding because it is too big to be allowed to fail. Major economies are shrinking rapidly or bouncing along at the bottom of an economic trough. Unemployment is rocketing, especially in manufacturing. And in some places, manufacturing is in free fall with business failure and fraud on the increase.

At the same time, corporations are more powerful than ever. Their executives are part of an insular, self-referential 'small world', and are paid at levels that many in society would see as 'obscene'. Divisions in society are greater than at any time in the last 40 years – a fact only too clearly illustrated by the fact that the family owning the USA retail corporation Walmart are paid more than the poorest 100 million people combined in that country. In many ways, this situation mirrors the excesses of the 'robber barons' of late nineteenth century USA capitalism, and represents an abrogation of the fiduciary responsibility – the responsibility of trust – that people and families at large place in them.

However, at present much economic geography is only partially engaged with these critical issues. In economic geography, windows on finance are few, and most are caught up with 'alterity' (Fuller et al. 2010) rather than with issues of securitisation and the finances of real businesses, including trade credit insurance. Financialization, which has figured prominently in economics, management, marketing and the business literature over the last decade, has only begun to filter into economic geography in recent years (see French et al. 2008), as Chandler's (1962; Chandler et al. 1997) three pillars of corporatism have been reduced to two, with 'production' having been externalised and corporate attention focussed increasingly instead on orchestrating revenue and finance streams (Taylor 1999). Institutionalist economic geography still sees inter-firm cooperation as the principal source of economic success, with the realities of brutal competition pushed into the background (Taylor and Plummer 2011).

At the same time, and as part of this same line of reasoning, firms are seen as not having boundaries. Instead, they are seen as 'fuzzily' merging one into another as they cooperate, collaborate and work towards some unspecified mutual good. In effect, an enterprise network is seen as being more important than its constituent enterprises. The knowledge sharing that is implicit in this interpretation of enterprise network relationships denies, to all intents and purposes, the overwhelming importance of intellectual property rights in present day business relationships. Intellectual property is a major source of profits in today's knowledge-based enterprises. Its protection is a major legal issue both nationally and internationally. But, for major sections of economic geography's interpretation of enterprise relationships money and knowledge flow unproblematically between firms to create innovations, skilled jobs, new firms and even global urban systems. At the global urban systems level, the economic geography approach is at its most naïve; simply counting the business headquarters to reveal the arithmetic of place-based success (see the critique in Taylor 2010). This functionalism is, however, far removed from the power plays and competitive brutality of global capitalism that the continuing recession of the current decade is making only too plain.

When these issues affecting large parts of current economic geography are brought together, it can be argued that sections of the discipline, especially institutionalist economic geography, are in danger of becoming both theoretically complacent and empirically vacuous. Theoretical complacency comes in the form of un-reconciled layers of theoretical ambiguity that has come from new theoretical constructs being added to earlier institutionalist ideas as has been outlined by Taylor (2006a, 2006b). New theories need to be tested and not blindly accepted. That economic geography is becoming empirically vacuous derives directly from theoretical complacency. The fact that sections of economic geography are devoted to, for example, LETS (local economic trading schemes) and CDFIs (Community Development Finance Initiatives) as alternatives to capitalism is akin to 'fiddling while Rome burns' in the economic turmoil of the current decade.

Realistically, can a bunch of locally linked and networked, small high-tech firms, producing niche products in an industry park with shared BBQ facilities, pull us out of recession? Or, can that only be done by a bunch of 'creatives' dreaming up consumerist ideas in the café cultures of central cities? These are the ideas offered by some sections of economic geography. Here, it is suggested that the discipline needs to come to grips with old-fashioned questions about the drivers of capitalism – the issues that have been 'othered' in the now-ended, cycle-denying, debt-fuelled boom of the 1990s and 2000s. To re-engage with the big issues of the day, economic geography needs to begin to re-explore the financial system, the meaning of manufacturing in a globalised world, IPR and inimitability, jurisdictional laxity, transaction governance and major labour market questions. The purpose of this chapter is to begin to broach some of these issues and to offer a possible foundation for the discipline to build its contribution to the understanding of and the future shaping of industrial transition.

**An Agenda for Action**

There are at least eight areas of economic geography that need to be re-addressed for the discipline to reconnect more fully with the dynamics of the current economic environment. The euphoria of the first years of the 21st century that led one British Prime Minister – Gordon Brown – to announce the end of economic boom and bust, has been brutally extinguished in the world's developed economies by a banking crisis (built on toxic bonds, securitised debt and insider trading) and a deepening sovereign debt crisis.

*Meaning of Manufacturing*

A first issue that needs to be re-examined is the importance and, in fact, *the meaning of manufacturing* in the developed economies of the 'west'. At the core of this issue is the increasingly complex and apparently ambiguous question: when is a manufactured product a service? Many engineered products, such as aircraft engines and pumps, for example, are now sold as 'years of service' rather than the sale of a physical product that the buyer will then own but must also pay to maintain, service and repair. However, as makers and designers of those engineered products strive to protect their intellectual property, and thus their competitive edge and their profits, they convert them, in effect, into embodied services. Add this development to increasing levels of productivity in manufacturing, and the productive and export base of countries like the UK falls from the public gaze (see the discussion in Bryson and Taylor 2006, Bryson et al. 2008).

At the same time, in countries like the UK, service sector employment has increased substantially and economic geographers have been at the forefront in the last 25 years in advocating that future economic success was to be found in the service economy: business and professional services, banking and financial services, retail and consumer services, and latterly in Europe, public sector services. Making things had become irrelevant. Outsourcing, the 'new' international division of labour, and the off-shoring of low-wage manufacturing jobs were all seen as the way forward. Indeed, at much the same time, sections of economic geography became transfixed by headcounts of service sector workers, creating a vision of world cities as the service and information hubs of a new world economic order – the undisputed orchestrators of economic growth.

In this increasingly inaccurate and distorted view of the dynamics of national and regional economies, the notion of 'basic' versus 'non-basic' sectors (Isard 1960) has been all but forgotten. It remains to be revived. Fifty years ago, basic sectors – principally manufacturing and production – were seen as creating new wealth in a place – building local economies and local prosperity. Non-basic sectors, like retailing, only redistributed the wealth within regions that had been created by those basic sectors. They had not created it. In the last two decades in the developed world, the creation and progressive easing of consumer credit has created the distorted view that retailing can create local economic growth: build a shopping centre and create local prosperity! Essentially, this is bank-created, credit-back delusion. Economic geographers need to unpack it.

*Global Race for Resources*

The issue of the basic/non-basic dichotomy in national and regional economies also leads into a second issue that economic geography needs to develop more fully and that is the *accelerating global race for resources*. The foundations of a production economy are the raw materials and recycled materials that are used to generate the intermediate and final manufacture of goods that societies demand. Without those resources and the goods made from them there are no foundations to support any demand for services. There is now more demand than ever for minerals and energy resources. There is a mining boom in places like Canada and Australia. Indeed, Australia has not experienced the recession and stagnation that has affected most of the developed world in the years since the 2008 financial crash. Now, the Norwegians are referred to as 'North Sea Sheiks', reflecting their oil and gas resources, and the Australians as 'the Saudis of Coal', for similar reasons. China and India are the leaders of a cohort of rapidly developing countries that are creating the demand for energy and resources to sustain the manufacturing that has been outsourced to them and to meet their own expanding domestic demand for goods.

This new surge in demand for energy and resources has major political consequences, especially in Africa, as countries like China seek resources they can control that are beyond the major mineral and energy corporations. The geopolitics of resources is an area of economic geography that the discipline needs to engage with more strongly. It brings with it a need to look more closely at transport and trade implications. Certainly, production is concentrated in China, East Asia and South Asia, but the nature of shipping is changing as vessels become larger but the navigable routes become more constrained as a result. At the same time, the costs of shipping are rising, with the result that in Europe some goods are now being sourced in Eastern Europe and Turkey. It is this nexus of issues around the supply, politics and transport of resources and energy that economic geography needs to engage with.

*Banking and the Financial System*

The geographical implications surrounding the supply of energy and resources together with the issues of basic and non-basic activity, and 'hidden' manufacturing in developed countries, lead into a third issue that economic geography needs to re-examine: the discipline needs a deeper and fuller *analysis of banking and the financial system*. To some extent this is beginning to happen with a new focus on the issue of financialization and the impact it has on the operations of listed companies and corporations. However, most economic geography is concerned principally with *the use of funds* in an economy, not with the *source of funds* (see the early work by Taylor and Thrift 1982). To get to the position of being able to anticipate what the local and regional impacts of the evolving financial system might be requires a deeper and spatially refined understanding of:

- How the financial system and its constituent parts function to cycle finance regionally, nationally and internationally and to support trade and enterprise;
- How deregulation and the commodification of money and debt have created and reinforced financial gaps that have always been evident in national and international systems (Taylor and Thrift 1982);
- How financial institutions and their overseers interpret their fiduciary responsibilities in an environment of deregulation;
- Why financial institutions can become too big too fail and can claim massive taxpayer bail-outs without the political or commercial retribution that would be meted out to other types of enterprise;
- How financial elites can remain impervious to and unaffected by their ineptitude.
- Even in a time of recession, economic geography remains fixated with the use of funds in local economies. The supply of funds is implicitly assumed to be unproblematic when patently it is not.

*Jurisdictional Laxity*

A fourth issue that economic geographers need to consider more deeply is the problem of *jurisdictional laxity*. Jurisdictional control of businesses, firms and enterprises is all too easily pushed to one side as decision-makers, like academics, are increasingly in thrall of the Friedmanesque vision of free markets: 'the market will decide'. But, for capitalism to function, legal underpinnings are essential to allow exchange and contractual relationships to exist in the commercial world beyond the level of barter. Jurisdictional controls, however, differ from place to place: an unavoidable consequence of the different paths towards economic development followed in different regions and nations that make up the world's global economic mosaic. Globalization has now thrown the issue of jurisdictional laxity to the fore as business owners and corporations seek to avoid the controls and the taxes imposed on them by the countries and regions within which they operate. What is emerging in the global economy is an inevitable 'race to the bottom' in terms of enterprise control. But this, however, is an issue that geographers seem reluctant to engage with.

There are at least three dimensions to the issue of jurisdictional laxity: corporate regulation, tax havens, and transfer pricing. Corporate regulation is a complex topic, but the literature suggests that, over time, corporate interests have fought to achieve, and have achieved, rights and privileges that make them ever more difficult to regulate. Some of the greatest insights have come from the USA where, through the nineteenth century, states fought one another to attract corporations through constitutional amendments and legislative acts that gave those corporations the freedom to design charters that suited their own interests most closely (Pendras 2011, Nader et al. 1976). For many years, this regulatory 'race to the bottom' was led by the State of New Jersey. But, in recent years, the State of Delaware has taken the lead, and a lead that is expected to grow (see Bebchuk and Hamdani 2002). Furthermore, according to Kolko (1963), Federal regulation in the USA has always been more in line with corporate interests than the public interest, as the 2008 global financial crisis showed very clearly. This lack of control and the erosion of standards it can involve have been made only too clear in the phone hacking scandal that envelops one of Delaware's more prominent corporations, News International, the heart of the Murdoch family empire.

Tax havens, too, are part of the corporations' strategies to circumvent regulation, to minimise their global tax burden. They have well established and well documented mechanisms such as transfer pricing that allow them to shuffle profits from where they are earned to the tax havens where they are lowest taxed. Just as with corporations, tax havens have now become the home of growing numbers of billionaire private equity owners seeking the very same goal. In short, we have corporations and the super rich that seek to expropriate as much wealth as possible from the countries in which they operate but who are unwilling to invest in the very communities from which they strip their incomes. But, countries too are part of this game. Tax havens are not only small Caribbean island states. Monaco is a tax haven within Europe like the island of Jersey or the Isle of Mann. And so, too, was Ireland before the financial crisis that started in 2008. Sadly, apart from a small number of now increasingly old studies, these issues of jurisdictional laxity remain largely opaque in the economic geography literature.

*Intellectual Property Rights*

A fifth issue that economic geography needs to address more closely is the economic significance of *intellectual property rights* (IPR) and the control of intellectual property itself. Currently, a large section of economic geography is underpinned by the somewhat naïve notion that ideas and information are freely exchanged between businesses and firms within enterprise networks. Much of this work has its foundations in research on Scandinavian firms where, it would seem, social conditions have built societies that tend to function in this way. In places like the UK and the USA, where capitalism functions rather more brutally, networks are exclusionary rather than inclusionary as the Scandinavian work would suggest. This brutality has been spelled out quite plainly in the USA context by Christopherson and Clark (2007), especially in the pharmaceuticals industry.

In the technology driven economies that are at the forefront of current economic development, the creation and control of intellectual property is pivotal, as the first issue on manufacturing discussed here has already suggested. Indeed, there are companies like ARM in the UK, engaged in hardware and software innovation in the personal communications field, that depend for their continued existence on their ability to trade and protect the intellectual property that they generate. If economic geographers are to engage more fully in developing an understanding of how intellectual property generation, trade and control affects local economic growth, there is a need to go beyond broad brush, general statements on 'information flows' and to come to grips with issues of information 'on what' and 'for what', and how it is traded between firms. It has to move beyond the implicit interpretation that all information is good and no information is bad.

*Transaction Governance*

Indeed, the issue of IPR leads into a sixth issue that economic geography needs to engage with more fully: the question of *transaction governance*. In significant sections of economic geography, transactions between business enterprises and firms are seen as occurring within networks and the power within those networks is seen as enacted only through the central, core enterprise(s). Transactions, however, in situations other than retail sales, are more complex. They are well recognized in the business and marketing literatures as involving legal constraints, contracts and power. They can involve unilateral relationships, where one partner in the transaction determines the functioning of that relationship, or they can be bilateral in their operation involving some degree of balance between the partners. Indeed, individual transactions can be a blend of both types of relationship depending on time and the dynamics between the partners (see the extensive discussion in Baker et al. 2002, Heide 1994, Jeffies and Reed 2000, Wathne and Heide 2000, Woolthuis et al. 2005). This is very much the situation within the Chinese 'gift economy' where magnanimity shown at one point in time may be used by a firm to exert power and control at a later time, and where power and control over firms is achieved by their incorporation into networks (Yang 1989). In short, the balance of power between partners in transactions may vary significantly over time. Consequently, if there are firms in a locality that are always or sometimes subordinated within transaction structures, as was suggested in the 1980s by Taylor and Thrift (1982, 1983) in their work on enterprise segmentation, then the dynamics and the economic future of that locality are going to be substantially shaped by those existing transaction governance arrangements.

*Social Elites*

Finally, two people-focused issues remain that economic geography needs to explore more fully in the workings of the current economic environment: the *role and significance of social elites*, along with the clubs and societies to which they belong, and, addressed further below, *labour market dynamics*, embracing unemployment and job creation and skills formation and training.

On *the role of elites* in the functioning of global corporations, there is a need to go beyond the emasculated notion of networks to revitalise C. Wright Mills' (1956) ideas on elites in the business world: *to explore the power of the empowered*. What is needed is a more nuanced understanding of the spatial concentration of corporate power that reflects the narrow, collusive control of wealth by the elites of society. Business elites are self-reinforcing, self-referential groups of the self-interested that operate within a 'small world' (Watts 1999) in which there is a high probability that any two people are connected by short paths of acquaintances (Davis et al. 2003). The small world of the corporate elite is a network of contagion (Conyon and Muldoon 2006). For Mills (1956), the power elite in society constitute cohesive interpersonal networks that embrace top business leaders, the executive branch of government, and the military that comprise 'a set of overlapping 'crowds' and intricately connected 'cliques'' (Mills 1956: 11) that are fundamental to the scope and retention of their power. They are grounded in common attitudes and values, social similarity and a distinctive world view of the upper and upper middle class who figuratively (and often literally in the European context) are 'the entitled'.

These networks are small. Davis et al. (2003) reported that the largest 516 US firms had only 4,538 directors. They have small inner circles of 'gatekeepers' with the result that they are relatively unaffected by major changes in membership. They have also been shown to be resilient to mergers and acquisitions, to the rising prominence of institutional investor, and to the greater public scrutiny of corporate board activity. That resilience exists for a number of reasons: the recruitment of people with shared backgrounds, the use of relational contracts to reinforce cohesion built on balancing the jeopardy of enforcement against the jeopardy of non-compliance, and the separation of ownership from control in large corporations. The result has been the marginalization of owners through the fragmentation of shareholdings, leaving the corporate elite beyond regulation and public scrutiny. After all, going back to Mills' (1956) ideas, the members of the

Financial Services Authority in the UK and Securities and Exchange Commission in the USA are drawn from the same elite that they are charged with regulating! There is in addition an apparent geographical dimension to the presence and persistence of elite networks that revolves around the working of exclusive upper-class clubs and also secret societies in major urban centres.

It remains for economic geography to begin to connect with these issues and to refine their geographical dimensions. This is particularly important since the three pillars of managerialism – production, marketing and management – have been reduced to two, with production increasingly being outsourced. It can be contended that the power of elites over marketing and management has, as a result, been strengthened (Taylor 1999).

*Labour Markets*

Finally, in the global economic shifts of the past decade, the *nature and dynamics of labour markets*, including skill formation, unemployment and job creation, have changed substantially. Focusing here on the world's developed economies, there are questions concerning training and apprenticeships, the erosion of the manufacturing skill base, the sustainability of public sector employment, retirement ages and pensions.

The issue raised here is straightforward. In the face of global economic pressures, and especially in the light of evidence from the UK and the USA, manufacturing that was made vulnerable from the 1960s onwards by the problems of Fordist mass production, malleable government and strong unions, has sought to remain competitive through the adoption of high value-added niche manufacturing strategies. However, the short-termism of business decision-making and business elites, coupled with an ageing manufacturing workforce, jobs that are now unattractive to the younger generation because they are manual and 'dirty', and inadequate skills training is rapidly eroding the manufacturing skill base. It means that firms' niche strategies to achieve growth, even before recession, are almost certainly doomed to fail, having now been 'eroded from below' (Taylor and Bryson 2010). Now, in the face of recession, and with every sign that financial and other service activities and employment will also contract and be more highly regulated than they have been in the past, the creation and maintenance of employment in the 'real economy' of manufacturing is going to be more important than ever. Economic geography needs to engage fully with this issue to inform policy and to gauge social attitudes on issues of education, training and skill to help to halt the 'erosion from below' that threatens manufacturing.

## Conclusion

National and regional economies, embedded in a rapidly changing global milieu, are under pressure to change: to rebuild, to restructure, to create jobs and to create growth that is sustainable into the future. It is important for economic geography to connect with this vital industrial transitioning. The empirical foundations of the discipline are important to counter the limited vision offered by free-market economics – the very vision that has brought deep recession to major sections of the global economy.

The eight sets of economic issues outlined in this chapter are major areas where economic geography can engage with and reinforce understandings of the local and regional impacts of the changing global economy and to inform policy to counteract recession and slow economic growth. Corporations have changed rapidly in the past 30 years, and the subordination of firms and enterprises within fragmented value chains has been, in some sectors, intense. The role of manufacturing in much of the developed world, other than Germany and Japan, has become obscured from the public gaze, devaluing its perceived economic importance compared with services. Large corporations and the social elites that control them have further changed the nature of enterprise, fostered by the commodification of money operated by a deregulated financial system, and cosseted by jurisdictional laxity. Transactional governance, linking firms into value chains, is in the process of change, and intellectual property rights that bolster and protect innovation in production and design is of ever-increasing importance.

However, recession in recent years and now slow and faltering recovery, that is having a major impact in Europe and the USA, serves to increase the urgency for the discipline of economic geography to fully engage with this raft of issues. In this way economic geography can retain and enhance its value for policy analysis and the development of understandings of how economic change impacts on regions, places and countries.

## References

Baker, G., Gibbons, R.and Murphy, K. 2002. Relational contracts and the theory of the firm. *Quarterly Journal of Economics*, 117, 39–84.

Bebchuk, L.A. and Hamdani, A. 2002. Vigorous race or leisurely walk: reconsidering the competition over corporate charters, *Yale Law Review*, 112, 553–616.

Bryson, J. and Taylor, M. 2006. *The Functioning Economic Geography of the West Midlands* (Summary Report and Main Report). The West Midlands Regional Observatory, Birmingham, UK.

Bryson, J., Taylor, M. and Cooper, R. 2008. Competing by Design, Specialization and Customization: Manufacturing Locks in the West Midlands (UK), *Geografiska Annaler*, 90(2), 173–86.

Chandler, A.D. 1962. *Structure and Strategy*. Cambridge, MA: Harvard University Press.

Chandler, A.D., Amatori, F. and Hikino, T. (eds) 1997. *Big Business and the Wealth of Nations*. Cambridge: Cambridge University Press.

Christopherson, S. and Clark, J. 2007. *Remaking Regional Economies: Power, Labor and Firm Strategies in the Knowledge Economy*. London and New York: Routledge.

Conyon, M.J. and Muldoon, M.R. 2006. The small world of corporate boards. *Journal of Business Finance and Accounting*, 33(9), 1321–43.

Davis, G.F., Yoo, M. and Baker, W.E. 2003. The small world of the American corporate elites, 1982–2001. *Strategic Organization*, 1(3), 301–26.

French, S., Leyshon, A. and Wainwright, T. 2008. *Financializing Space and Financialization*. Paper presented to the ESRC Financialization and Competitiveness seminar, Northumbria University, UK.

Fuller D., Jonas A. and Lee R. (eds) 2010. *Interrogating Alterity*. Farnham: Ashgate.

Heide, J.B. 1994. Interorganizational governance in marketing channels. *Journal of Marketing*, 58, 74–85.

Isard, W. 1960. *Methods of Regional Analysis*. Cambridge, MA: MIT Press.

Jeffries, F.L. and Reed, R. 2000. Trust and adaptation in relational contracts. *Academy of Management Review*, 25(4), 873–82.

Kolko, G. 1963. *The triumph of conservatism*. New York: The Free Press.

Mills, C.W. 1956. The Power Elite. New York: Oxford Univ. Press.

Nader, R., Green, M. and Seligman, J. 1976. *Taming the Giant Corporation*. New York: Norton.

Pendras, M. 2011. Law and the political geography of US corporate regulation. *Space and Polity*, 15(1), 1–20.

Taylor, M. 1999. The Dynamics of US Managerialism and American Corporations', in *The American Century: Consensus and Coercion in the Projection of American Power,* edited by D. Slater and P. Taylor. Oxford: Blackwell, 51–66.

Taylor, M. 2006a. Fragments and Gaps: Exploring the Theory of the Firm, in *Understanding the Firm: Spatial and Organizational Dimensions*, edited by M. Taylor and P. Oinas. Oxford: Oxford University Press, 3–31.

Taylor, M. 2006b. The Firm: Coalitions, Communities and Collective Agency, in *Understanding the Firm: Spatial and Organizational Dimensions*, edited by M. Taylor and P. Oinas. Oxford: Oxford University Press, 87–116.

Taylor, M. 2010. Clusters: a mesmerising mantra. *Tijdschrift voor Economische en Sociale Geografie*, 101(3), 276–86.

Taylor, M. and Bryson, J. 2010. Erosion from above, erosion from below: labour, value chain relegation and manufacturing sustainability, in *Missing Links in Local Labour Geography*, edited by A. Bergene et al. Farnham: Ashgate, 75–93.

Taylor, M. and Plummer, P. 2011. Endogenous Regional Theory: A Geographer's Perspective and Interpretation, in *Endogenous Regional Development*, edited by R. Stimson et al. Cheltenham: Edward Elgar, 39–48.

Taylor, M. and Thrift, N. 1982. Industrial linkage and the segmented economy: 1. Some theoretical proposals. *Environment and Planning A*, 14, 1601–13.

Taylor, M. and Thrift, N. 1983. Business organisation segmentation and location. *Regional Studies*, 17(6), 445–65.

Wathne, K.H. and Heide, J.B. 2000. Opportunism in inter-firm relationships: forms, outcomes and solutions. *Journal of Marketing*, 64, 36–51.

Watts, R. 1999. Networks, dynamics and the small-world phenomenon. *American Journal of Sociology*, 105, 493–527.

Woolthuis, R., Hillebrande, B. and Nooteboom, B. 2005. Trusts, Contracts and Relationship Development. *Organization Studies*, 26, 813–40.

Yang, M. 1989. The gift economy and state power in China. *Comparative Studies in Society and History*, 31(1), 25–54.

Chapter 3

# The Transition of the Automotive Industry as a Catalyst for Cross-Border Networking? The Case of the Greater Region SaarLorLux

Christoph K. Hahn

## Introduction

The automotive industry is about to enter an era of fundamental change. General trends, such as the globalization of value chains, major technological innovations, and the global depression in 2008–09 have consequences on many regions and enterprises in the traditional bases of the value chain as well as in the growing markets (Maxton and Wormald 2004, Jürgens and Krzywdzinski 2009).

The ongoing transition process can be divided into two major strands. On the one hand, profound change occurs in the structural and spatial organization of the industry. Since the implementation of the concept of lean production by Toyota, Honda, and Nissan during the 1970s and 1980s, automobile producers worldwide have been increasingly concentrating on the final assembly of the car while delegating the construction of most modules to their first tier suppliers (Maxton and Wormald 2004). At the same time, the industry has been installing more and more global linkages and networks. Today, large parts of the industry are integrated into global value chains or are challenged by global competition (Sturgeon, Van Biesebroeck and Gereffi 2008). Smaller components as well as modules, but also design and R&D expertise, are increasingly sourced at the global level.

On the other hand, the performance of the automotive industry is influenced by the need for technological innovations, such as alternative propulsion technologies or lightweight materials (Maxton and Wormald 2004). Enterprises, which until today focus on traditional propulsion technologies, will have to diversify their portfolio. However, at the moment, uncertainty remains concerning which specific technology will be used in the future and when. This transition could prove particularly difficult for small and medium sized suppliers as their limited resources do not enable them to focus simultaneously on several trajectories.

This uncertainty and the growing competition due to the globalization of the value chain jeopardize the existence of many subordinate suppliers in high-cost regions. Recently, the consequences for these enterprises were intensified as a result of the global recession in 2008–09. In the course of the economic slump, many first tier suppliers followed a strategy of again re-integrating numerous

activities which had initially been conducted in-house, but then outsourced to
other regional suppliers. This caused a severe decline in orders to subordinate
suppliers. At the same time, it is extremely hard for them to build up relations with
new clients due to their weak position in the automotive supply chain and their
limited resources.

Enterprises, regional politicians, and trade associations respond to these threats,
among others, by the installation of cluster initiatives and innovation networks.[1]
These initiatives are meant to serve as a mediator for inter-firm cooperation. In the
special case of the transnational Greater Region SaarLorLux, an additional focus
is placed on the creation of synergies within the entire region by furthering cross-
border contacts, particularly joint R&D projects (Staatskanzlei Saarland 2003).
These efforts are based on the fact that all parts of the Greater Region have a large
number of enterprises and research institutes working for the automotive industry.

Thus, at first glance, it seems as if network initiatives serve as an efficient response
to the current needs of the automotive industry and as an aid to enterprises adapting
to the changing environment. Being a member of one of the automotive networks
could be beneficial for a company as this enhances their chances of acquiring new
clients and installing joint innovation projects. Furthermore, implementing a cross-
border strategy enables an enterprise to profit from complementary skills which are
provided by foreign companies and research institutes.

The aim of this chapter is to verify whether such networking strategies really
characterize the process of adaptation within the automotive industry of the Greater
Region SaarLorLux. In pursuing this aim, several questions are addressed. First,
does the transition of the automotive industry lead to cross-border networking
within the Greater Region SaarLorLux? Second, what prospects appear to be
supportive? Third, what challenges slow down such approaches?

The first section presents a conceptual framework of regional innovation
networks and cluster initiatives. This is followed by an introduction to the
Greater Region SaarLorLux and its automotive industry in section two. The third
section presents the results of a survey, which examined the performance of 77
automotive enterprises within the Greater Region. This survey focused on the
enterprises' perception of prospects and challenges for cross-border economic
activity. The findings of these analyses lead to some conclusive remarks in the
final section of the chapter.

---

1    The terms network and cluster initiatives will be used interchangeably. They do
not necessarily confirm to their scientific definitions (Martin and Sunley 2003), as the
terms represent the political view and the labels of the initiatives within the Greater Region
SaarLorLux.

## Regional Innovation Networks and Cluster Initiatives

Following the concepts of lean production and flexible specialization, economic activities are increasingly organized in inter-firm relations rather than in-house. Within the automotive industry, for example, the automobile assemblers' share of added value is constantly decreasing, as more and more parts are produced by suppliers (Maxton and Wormald 2004, VDA 2004, Sturgeon, Van Biesebroeck and Gereffi 2008).

Parallel to the growing relevance of inter-firm relations, scholars within economic geography have started to integrate a relational perspective into their research agenda (Bathelt 2006). This approach is a response to the disregard of the active role of economic agents by traditional views. Instead of treating regions as actors and using spatial variables for explanation, the relational perspective focuses on the people who act in enterprises and interact with their environment. This view takes into account that every agent is embedded in its social and institutional setting, which influences their decisions and actions. To consider these facets of economic agents and their performance, studies in economic geography need to adopt a micro-level approach based on extensive empirical work. In doing so, the relational view also aims at better analyzing patterns of local response to global processes.

Numerous studies have focused in particular on the interactive dimension of the innovation process (Torre and Rallet 2005, Boschma and Frenken 2009). This chapter argues that the study of innovation deserves special attention within economic studies, as innovation activity is essential for the long-term competitiveness of an enterprise. The work of Torre and Rallet (2005) and Boschma and Frenken (2009) highlight the influence of various forms of proximity on the innovation process. On one hand, geographical proximity between the partners of inter-firm R&D activities seems to have a positive effect on the innovation process, as it eases frequent face-to-face meetings and the exchange of tacit knowledge. However, geographical proximity alone is neither a necessary nor a sufficient condition for joint R&D projects. Furthermore, too much geographical proximity or focusing too much on local partners could have negative effects, including increased risk of lock-in and the missing out on new technological or organizational innovations (Maskell and Malmberg 2007).

Consequently, other aspects of proximity and distance are needed, including social, organizational, cognitive, relational, or institutional (Coenen, Moodysson and Asheim 2004, Boschma and Frenken 2009). As these dimensions are hard to quantify, the terms *qualitative* or *soft* will be used in this chapter. In contrast to geographical proximity, these dimensions describe the more intangible environment in which innovation in particular, and economic action in general, takes place. If the members of a joint R&D project share a common institutional and jurisdictional setting, if they are embedded in similar organizational structures, and if their knowledge bases are compatible, the agents will be close to one another regarding these soft forms of proximity. But once again, too much proximity might be detrimental as it increases the risk of lock-in. Thus, only when all dimensions of proximity are balanced, will they help to minimize transaction costs and will therefore be able to serve as a mediator or infrastructure fostering beneficial contacts between agents (Lundquist and Trippl 2009, Trippl 2010). Such a favorable state might also be 'artificially' created, as the soft aspects of the environment are not stable, but actively shaped and re-shaped (Bathelt 2006).

Companies, regional politicians, and trade associations have adopted this idea and tried to create the necessary infrastructure and a supportive setting via regional cluster or network initiatives. Such policies aim, in general, to enhance the competitive position of the regional enterprises. In this context, Andersson et al. (2004) identify five domains of cluster policies which help to support the regional economy:

- Broker policies aimed at facilitating interaction at the regional level by furthering contact and exchange among enterprises. Therefore, platforms for dialogue, such as joint meetings and factory tours, and technology transfer between the companies are arranged.
- Demand side policies, such as public procurement or the provision of information and data bases, might increase the economic position of regional companies.
- Training, which is in particular essential for small and medium sized enterprises (SMEs), upgrades the human capital.
- Promotion of external linkages aimed at reducing the risk of positive feedback and lock-in.
- Improvements in broader framework conditions regarding the physical, institutional and judicial infrastructure serve as a facilitator and catalyst for the other four dimensions.

This list stresses the rather systemic character of cluster policies, which emphasize the improvement of 'soft' infrastructures. Consequently, most initiatives are not addressed within single companies but to the entire value chain or even a group of related value chains. Nonetheless, they are also meant to have positive effects at the level of the enterprise.

Regional clusters might become part of regional innovation systems if they were supported by financial, administrative, and research infrastructures (Cooke 2001). Such an innovation system further stabilizes the economic position of the enterprises and augments the effects of cluster initiatives, as it particularly strengthens the innovative ability of the companies. The more responsibility that is delegated to the regional level, the more likely a region is to form such a regional innovation system. This will be the case if, for example, regionally available private capital, regional public budgets, and regional university-industry strategies exist. Furthermore, regional innovation systems require entrepreneurial competence to make strategic decisions within the region. An environment of independent companies, headquarters, and R&D departments meets these requirements and facilitates the creation of a regional innovation system.

Until today, most research on cluster initiatives and regional innovation systems has been conducted in regions which belong to only one nation state (for exceptions, see Coenen, Moodysson and Asheim 2004, Lundquist and Trippl 2009, Trippl 2010). Within the European Union, however, more and more regions along national borders merge, at least officially, into cross-border regions (Perkmann 2003). As these are often peripheral and economically weak regions, the demand for effective policies to improve the economic power is notably high. Consequently, adopting the concepts of regional innovation systems and clusters to the cross-border context seems promising and, in principle, possible (Coenen, Moodysson and Asheim 2004, Trippl 2010). Also the European Commission seems to be convinced by such efforts, as it supports cross-border initiatives within its INTERREG- and European-Territorial-Cooperation programs. However, the success of cross-border networking as a catalyst for more intensive cross-border economic integration, and in particular innovation activities, often remains questionable as numerous barriers are still to be removed (Coenen, Moodysson and Asheim 2004, Trippl 2010).

**The Greater Region SaarLorLux and its Automotive Industry**

The Greater Region SaarLorLux is a union of five administrative jurisdictions spreading across the four nation states of Belgium, France, Germany and Luxembourg. This constellation is the result of increasing cross-border cooperation within the border zone over the last 45 years. In the 1960s, the administrations of the French region of Lorraine and the German state of Saarland began informal discussions on trans-boundary cooperation. These initiatives were institutionalized in 1971 within a cross-border Regional Commission, which also included Luxembourg and parts of the German state of Rhineland-Palatinate. The Greater Region SaarLorLux in its present shape was formed in 1995, further encompassing the entire German state of Rhineland-Palatinate, the Belgian region of Wallonia as well as the French- and German-speaking Communities of Belgium (Dörrenbächer 2010). Since this time, various efforts have been taken to create a common identity within the Greater Region SaarLorLux, particularly based on the strategy paper *Zukunftsbild 2020* (Staatskanzlei Saarland 2003). Although this aim is far from being achieved, the region constitutes, at least in parts, an integrated functional area. It has, for example, a high number of cross-border commuters and relatively high levels of cross-border foreign direct investments (Dörrenbächer and Schulz 2005). Other examples of the integration process are recent cross-border cluster activities, such as the formation of the tri-regional cluster initiative 'INTERMAT' in the materials science industry of Lorraine, Luxembourg and Wallonia.

Among others, this merging process benefits from a shared history, particularly regarding a comparable economic development path in large parts of the Greater

Region SaarLorLux. Lorraine, Luxembourg, Saarland and Wallonia were dominated by iron ore and coal mining as well as iron and steel industries until the 1970s. However, the decline of these traditional industries since the 1960s resulted in tremendous job losses (Dörrenbächer and Schulz 2005).

To counteract these developments, regional and national policies were enacted to support the automotive sector in playing a major role in the restructuring of the regional economies. The establishment of a Ford plant in Saarlouis (Saarland) in 1966 was a result of such locational policies. In Lorraine, the formation of the automotive industry gained momentum around 1980 by Peugeot-Citroën's launching of an engine plant and the production of drive trains next to Metz. In the past decades, several enterprises have started businesses within the region or have shifted their focus away from the declining iron and steel industries toward the automotive industry.

Today, official statistics identify about 220 companies within the Greater Region SaarLorLux which are directly integrated into the automotive production process.[2] Most of them are members of regional cluster initiatives, which were established through public initiatives in each sub-region within the last 15 years. According to these networks, the automotive industry of the Greater Region consists of even more firms: over 650 companies, employing approximately 160,000 people. These numbers differ from official statistics, as the networks also take the subordinate suppliers into account. The aim of the initiatives is to foster regional cooperation, technology transfer, and marketing. Therefore, these networks provide databases with information on the member companies. They also organize meetings, seminars, provide factory tours, and represent the regional automotive industry at trade fairs. Recently, politicians and the five automobile networks announced a number of smaller initiatives to establish a common marketing strategy in order to be more globally competitive. Furthermore, access to research institutes across the border will be facilitated in order to create synergies and improve the potential profits throughout the entire Greater Region SaarLorLux.

## Range of Relations within the Automotive Industry of the Greater Region SaarLorLux

The various forms of relations within the automotive industry are examined through a two-tiered approach. First, a postal survey was administered to all 674 enterprises listed in the databases of the cluster initiatives. Second, the performances of 15 companies were analyzed in greater detail through guided interviews. This method, which is comparable to several studies of the *Groupe de Recherche Européen sur les Milieux Innovateurs* (GREMI) (Aydalot 1986), was the most effective method as the size of the population did not allow for social network analysis due to lack of availability of complete network data.

---

2   As Wallonia does not provide updated data, it is not included in this figure.

Between November 2009 and March 2010, 77 enterprises (about 11 per cent) participated in the postal survey by sending back the completed questionnaire. Some organizational features of these companies are summarized in Table 3.1.

**Table 3.1     Organizational characteristics of the surveyed enterprises and cooperation within the Greater Region SaarLorLux**

|  | Number of enterprises |
|---|---|
| Location: | |
|     Lorraine | 26 (260)* |
|     Luxembourg | 4 (17) |
|     Rhineland-Palatinate | 13 (150) |
|     Saarland | 26 (190) |
|     Wallonia | 8 (57) |
| | |
| Size in terms of number of employees: | |
|     20 or less | 22 (195) |
|     21 to 250 | 40 (371) |
|     More than 250 | 15 (108) |
| | |
| Legal status: | |
|     Independent enterprise | 40 |
|     Subsidiary | 37 |
| | |
| Cooperation within the automotive industry (with partners located in the Greater Region SaarLorLux): | 40 |
|     Yes | 12 |
|     No | 25 |
|     No data / no cooperation at all | |

\* *Note*: Figures in brackets indicate the basic population.

*Source*: Survey Results

As indicated in Table 3.1, more than half of the enterprises cooperate with partners who are also located in the Greater Region SaarLorLux. Thus, at first glance, the region seems to form a common functional area. However, the degree and relevance of regional integration varies according to the position of the enterprise within the automotive value chain. Whereas 40 per cent of the subordinate suppliers cooperate only within their regional context, eight out of nine original equipment manufacturers (OEMs) and first tier suppliers also have partners who are located outside the region.

In addition to a particular company's position in the pyramid of suppliers, the legal status of the respective enterprise influences their level of embeddedness. That is, subsidiaries have limited authority to choose partners, but rather, rely on the decisions of the holding company. Consequently, these subsidiaries tend not to engage in regional networking as a way to search for new suppliers or clients, as evidenced in the following quotes:

> At our production plant we cannot choose our suppliers, partners, or clients. The central buying office gives us a list with the respective enterprises which are located all over Europe ... It may happen that we receive the order to stop cooperation with some of our partners within one or two weeks, whatever the reason may be. (Plant Assistant of a subsidiary of a first tier supplier, translation by the author)
>
> The only client we serve at our site is Ford Saarlouis. I would not say that it would cause trouble if I asked the holding company for the permission to allow for further regional cooperation—but this is not wanted. Instead, the headquarters makes administrative and strategic decisions. Our task is to produce for Ford—that's it! (Production supervisor of a subsidiary of a first tier supplier, translation by the author)

On the other hand, independent enterprises have the authority to use regional networks, meetings with companies located in the region, or even contacts established during former employment to foster new relations. Two comments illustrate this point:

> Thanks to the contacts we established during our former employment positions in a first tier, we had a large pool of potential clients to contact after starting our own enterprise ... Another major strategy has been to get in touch with other enterprises at regional meetings, as face-to-face contacts are of paramount importance concerning the services we offer.
>
> (Director of an independent second tier supplier, translation by the author)
>
> If one of our clients is asking for a special job which we cannot handle, we recommend the interested party to one of our partners in the region who has the authority to deal with that job. As this partner acts in the same way, both companies benefit from this relationship, as we can easily generate more jobs.
>
> (Sales manager of an independent second tier supplier, translation by the author)

In general, cooperation tends to be established in the rather knowledge and trust extensive domain of production (34 out of 40 enterprises), whereas joint R&D (14 out of 40) or marketing (5 out of 40) activities are exceptional. Consequently, most relationships can be characterized as classical buyer-supplier relationships between the subordinate suppliers and the OEMs and large first tier suppliers.

Some of the reasons for the low proportion of trust intensive relationships are discussed later in this chapter. First, however, the effect of the borders on cooperation is examined in greater detail.

*Cross-border Cooperation within the Greater Region SaarLorLux*

The level of cross-border cooperation (eleven companies) is relatively low. Rather, most of the enterprises cooperating within the Greater Region SaarLorLux are regionally embedded in the narrowest sense. That is, partners, clients and suppliers are located primarily within a respective region, and to a lesser degree within the respective nation state, rather than in foreign parts of the Greater Region. Therefore, the automotive industry within the Greater Region SaarLorLux appears more likely to form a conglomerate of five separate regions instead of one integrated functional region. Results indicated that it was only between the two German sub-regions that cooperation is more common (i.e. eight relationships between enterprises from Rhineland-Palatinate and Saarland).

To identify the reasons for the given regional focus of the enterprises, the interrelationship of proximity and innovation serves as a starting point. As mentioned above, geographical as well as soft forms of proximity will stimulate innovation if they are well balanced. These findings are, however, not limited to the process of innovation, but are also applicable to analyzing other forms of cooperation, as all kinds of inter-firm relations are supported by a certain degree of proximity.

In addition to the common differentiation between geographical proximity on one hand and qualitative forms on the other, the subsequent evaluation further draws upon the insights gained by Coenen, Moodysson and Asheim (2004). In their work on proximities within the medical sciences of the Danish-Swedish border zone they broaden the concept of geographical proximity. Instead of measuring only distance, the availability of transport and communication infrastructure is also taken into account. Coenen, Moodysson and Asheim (2004) call this *functional proximity,* which corresponds to accessibility rather than distance. Supported by their empirical findings, they conclude that functional proximity is a necessary, though not sufficient condition for interactive innovation.

Within the automotive industry of the Greater Region SaarLorLux, 34 out of 72 enterprises state that the absence of trans-border platforms to get in touch with each other or to exchange information is a main barrier for cross-border economic activity. The same number of companies indicated lacking adequate information about the structure of the automotive industry in the other sub-regions, as reflected in the following comments:

> We simply do not know if there are potential partners located in Lorraine or Luxembourg. And I assume that the French or Luxembourgian companies also

lack information about the structure of Saarland's economy … And I do not see a promising way how we can establish fruitful contacts with French or Luxembourgian partners.

(Director of an independent second tier supplier in Saarland, translation by the author)

The network init'ative of the Saarland recently published a map showing the location and type of the enterprises belonging to Saarland's automotive sector. If such a map were drawn for the Greater Region, with information added in English, French and German, and if international meetings were arranged, this would help us build relations across the border—and we would like to do so.

(Plant manager of an independent second tier supplier, translation by the author)

These statements make clear that the degree of cross-border functional proximity within the automotive industry of the Greater Region SaarLorLux is low. Access to enterprises located in foreign sub-regions is hindered due to a lack of cross-border communication infrastructure. In particular, many SMEs suffer from this situation, as their limited workforce does not enable them to do intensive cross-border acquisition independently.

Following the statement of the plant manager cited above and Morgan's studies on the Welsh Development Agency and the *Learning Region* (Morgan 1997), public institutions, including the cluster initiatives, have, in principle, shown the competence to establish the desired support infrastructure within the entire Greater Region. Furthermore, the cross-border programs of the European Commission aim to provide financial support for such activities. However, apart from smaller temporary events, intensive cooperation among the regional cluster initiatives has been mainly non-existent until recently.

One reason for this situation might be the uneven distribution of power within the cluster initiatives. Large stakeholders with a lot of influence on the activities organized by the initiatives appear to have no interest in supporting efforts toward cross-border networks. A member of the steering group of one network stated: 'I think the big ones do not want to support their suppliers in searching for other clients or entering foreign markets, as this would only mean increased competition for the big ones'. On the other hand, the multitude of SMEs, which would benefit from cross-border initiatives, is only in a weak position within the networks. Thus, as long as the large stakeholders continue to possess that much power, cross-border cluster initiatives are unlikely to be installed and the cross-border functional proximity will not be fully optimized. Another obstacle relates to conflicting interests among regional policies within the Greater Region. Recently, an INTERREG-IVa project proposal which aimed at fostering cross-border contacts within the automotive industry failed, as representatives of the five sub-regions did not approve it unanimously.

However, even if the functional proximity were increased through integration of the five initiatives, further challenges, including those in the social domain of proximity, would remain. For example, 26 enterprises, primarily SMEs, consider that the different languages spoken within the Greater Region constitute a barrier. Their small staff minimizes chances of employees speaking French and German fluently. Thus, few SMEs are able to contact foreign companies. Another barrier for cross-border cooperation is the lack of authority to choose partners or suppliers independently which is typical for many subsidiaries. As noted in one interview:

> We are a subsidiary in Lorraine belonging to a German holding company ... If I tried to contact, for example, a department at Saarland University to establish a partnership to optimize our production process, the holding company would say: 'Stop, Germany is our field of activity.' I can search for contacts in France, but not across the border.
>
> (Plant manager of a subsidiary of a third tier supplier, translation by the author)

Another barrier relates to the existence of different legal systems within the four nation states. Several interviewees, again primarily in SMEs, stated that they do not know how to arrange contracts with foreign clients or partners, as such contracts differ from those that they commonly use. Consequently, they would appreciate obtaining assistance by the cluster initiatives or trade associations. Finally, mental barriers hamper cross-border cooperation. Several interviewees mentioned that there are prejudices against the performance of foreign enterprises.

Once more, at least some of these barriers could be, in principle, removed if the cluster initiatives, trade associations, and public institutions increased their focus on a cross-border approach. In particular, the facilitation of networks, the provision of information on the economic structure of the other sub-regions, and the organization of cross-border events such as business-matchings appear to be promising strategies. Furthermore, language courses and seminars on international contract law would assist in preparing the enterprises for cross-border economic activities.

*Influence of the Automotive Industry's Transition on Cross-border Networking*

As outlined in the introduction, the automotive industry is about to enter an era of radical technological innovation. The necessity for such innovation results from among other things, tightened ecological restrictions within the context of global change. That is, a primary aim is to reduce automobile-based $CO_2$-emissions. Maxton and Wormald (2004) identify several strategies to meet this goal:

- First, they point out that the existing engine and fuelling technologies are not yet maxed out. Instead, improvements can still be achieved, for example by an optimized automatic gearbox or high-pressure direct-injection diesel engines. Recently, the Greater Region SaarLorLux plays a significant part in such developments. ZF Getriebe GmbH is about to extend the production of its eight-gear automatic transmission in Saarbrücken (Saarland) and Bosch produces major elements of the direct-injection system, Common Rail, in Homburg (Saarland).
- Another step is to reduce the mass of the car, mainly by replacing steel with aluminum. Recently, the share of aluminum in the total weight of the car has been increased through modifications such as using aluminum cylinder blocks for engines. However, little progress can be observed in the use of aluminum chassis or body structures due to safety concerns. Another obstacle relates to the high cost of producing aluminum.
- A radical alternative to reduce the $CO_2$-emissions of the car is to minimize and even stop the use of conventional engines and to replace oil with other energy sources. Hybrid technology is currently the most advanced alternative with Toyota the company being the first to bring it to the mass market. However, technologies which substitute fossil fuels completely are not yet mature enough for mass production. Electronic cars in particular, face the challenges of efficient battery storage and range. Furthermore, they will not necessarily reduce overall emissions if they are not based on renewable energy sources. On the other hand, cars using hydrogen fuel cells have to overcome barriers of fuelling infrastructure, cost, reliability, and durability before they can be considered as alternatives for mass markets.

These statements clarify that the automotive industry is in transition. The beginnings of another technological revolution have already had effects on automobile assemblers and parts of the supply industry. The impacts will increase as soon as radically new propulsion technologies and lightweight materials are ready for mass production, as this will result in modifications of almost all modules of the car. In the transition, first tier suppliers will play an important part in the development processes, as more and more innovation activity is transferred from the OEMs to their suppliers or is organized in joint projects (VDA 2004, Maxton and Wormald 2004). Furthermore, in order to meet global competition, subordinate suppliers will continue to be forced to develop new products, production processes, and relationships within the automotive sector.

At first glance, the automotive sector in the Greater Region SaarLorLux might have the potential to contribute to a significant share in the innovation process, particularly if the cross-border functional proximity were optimized. On one hand, leading first tier suppliers are located in the region (Figure 3.1). On the other hand, more than 80 per cent of the companies belong to the group of SMEs, which are seen to be flexible and innovative. The region's status is further supported by its large number of enterprises and an impressive diversity of skills in the labor force.

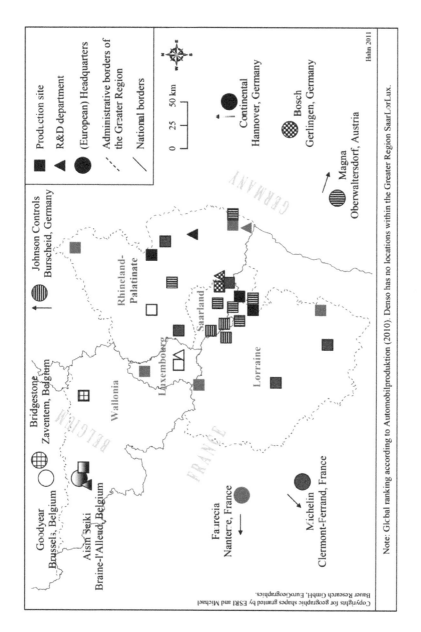

Figure 3.1    Locations of global top ten automotive suppliers within the Greater Region SaarLorLux

The cluster initiatives identified in particular the topic of light-weight materials as a promising domain of innovation activity within the regional automotive industry. Besides a large number of enterprises working in this field, several research institutes within the region already focus on material sciences, including the Leibniz Institute for New Materials, situated in Saarbrücken, the Materia Nova Materials R&D Centre, located in Mons (Wallonia), or the Institute Jean Lamour in Nancy (Lorraine). Lastly, the formation of the new network INTERMAT in Lorraine, Luxembourg and Wallonia provides potential links for cross-sectoral R&D projects.

Although the transition process of the automotive industry brings about some promising opportunities to the Greater Region SaarLorLux, several arguments contradict these prospects. Firstly, a very basic prerequisite of interactive innovation is mainly absent. As early works on inter-firm cooperation show, joint R&D activities require mutual trust and understanding (Håkansson and Johanson 1993, Malmberg and Maskell 1997). As these studies stress further, it usually takes time to build trust and understanding. However, as stated in many interviews, the relationships between the numerous regional second and third tier suppliers and the OEMs and first tier suppliers are seldom of a reliable and long-lasting character. As already shown here, this does not inhibit cooperation at the regional level in general. Rather, the consequence is that most relationships between OEMs and first tier suppliers on one hand and subordinate suppliers on the other lack high levels of trust and informal relations. Therefore, most of the cooperation at the regional level consists of short-term buyer—supplier relations rather than joint R&D or marketing initiatives.

This fact is further supported by a second barrier toward joint innovation projects, which is the legal status of the OEMs and first tier suppliers within the Greater Region SaarLorLux. They are mainly subsidiaries focusing on production, whereas most of the headquarters and R&D departments are located outside the region (Figure 3.1). The example of the eight-gear automatic transmission by ZF Getriebe GmbH illustrates the spatial organization of the innovation process. As the R&D department is located outside the Greater Region SaarLorLux, about 250 kilometers from the production site in Saarbrücken, other enterprises from within the Greater Region were not involved in innovation activities. Instead, ZF Getriebe GmbH cooperated with companies that were situated close to their R&D department. The resulting structure seems to be typical for peripheral locations within the automotive industry (Bilbao-Ubillos 2010), and is comparable to Markusen's (1996) *Hub-and-Spoke District* and *Satellite Platform District* which hardly generated innovation at the regional level. The main characteristics of such a structure are as follows:

- The regional economy is heavily dependent on large, focal enterprises such as the OEMs and first tier suppliers.
- These companies maintain primarily subsidiaries within the region, so that strategic decision-making and innovation takes place outside the region.

- Large enterprises are embedded at a supra-regional level, as they maintain substantial links to clients, suppliers, and partners external to the region.
- Small enterprises are mainly embedded at the regional level, as relations to external companies are rare.

To enable a comprehensive mobilization of regionally bounded knowledge, links between small suppliers and the R&D departments of first tiers and OEMs would have to be installed, which is not an easy task. On one hand, small enterprises do not necessarily possess the resources and self-confidence to contact the R&D department of a large first tier supplier located outside the Greater Region SaarLorLux. Having said this, even if the cluster initiatives were taking an active part in promoting relationships between regional suppliers and large first tier suppliers or OEMs, success would not be guaranteed. As Tödtling and Trippl (2004) state, many of these global players have little interest in becoming engaged in a process of social networking on the regional level. Thus, the organizational dimension of the enterprise does necessarily contribute to the creation of a regional innovation system within the automotive industry of the Greater Region.

A third obstacle relates to a lack of comprehensive university—industry strategies, at regional as well as at a cross-border level. Similar to the findings of Rutherford and Holmes (2008) who analyze university—industry networks within the Southern Ontario automotive clusters, existing strategies seem to disregard the needs of the majority of SMEs. Consequently, cooperation between SMEs and research institutes is rare, whereas joint R&D projects between research institutes and OEMs or first tier suppliers are thoroughly common. Furthermore, interviewees in SMEs mentioned that they perceive themselves as being too marginal in order to attract the interest of the research institutes. This perception is further intensified in the case of cross-border cooperation, as linguistic barriers increase the distance to foreign institutes.

All of these circumstances constitute an environment which makes it unlikely that the development of new technologies within the automotive industry will take place in the Greater Region SaarLorLux, particularly not at a cross-border level. Instead of being a catalyst for cross-border innovation networks, the transition process of the industry appears to have had a cathartic effect. Numerous interviewees, mainly in subordinate suppliers, stated that, due to growing competition and the economic depression in 2008–09, they are trying to diversify their respective company's activities. Several tool and machinery manufacturers, for example, perceive medical technology as a promising alternative for their business. However, at the same time they mentioned difficulties in starting such new activities, as they do not possess contacts with these industries.

An option which will be beneficial to broader parts of the automobile production chain will thus be the evolution of *trans-border* and *trans-sectoral* networks by public initiatives (Martin and Sunley 2003, Tödtling and Trippl 2004). As illustrated in this chapter, several enterprises are willing to enter either foreign markets (within the Greater Region SaarLorLux) or shift to new industrial sectors.

Thus, regional cluster policies could support such efforts by interlinking both the independent regional initiatives among themselves and with other industries, leading to partnerships which do not necessarily focus on the automotive sector. However, as the following statement shows, various lobbyist interests seem to disapprove of such developments:

In the course of the recent depression, some enterprises told us that they would like to start businesses in other sectors. However, our network takes care of the automotive supply chain. Thus, we have little interest in supporting their diversification process, as this is a loss for our supply chain. (Contact person of a cluster initiative, translation by the author)

**Conclusion**

The automotive transition process puts high pressure on the enterprises within the Greater Region SaarLorLux, especially on the small and medium sized subordinate suppliers. Growing global competition, the need to design new technologies, and the global depression made companies develop various strategies. Some concentrated on their core business, while others favored a diversification process, looking for alternative activities. Some increased their R&D expenditures, others wished to cross the border in order to profit from a larger pool of potential contacts and additional skills provided by foreign research institutes. Consequently, the transition of the automotive industry supports, among others, the creation of cross-border networking. However, the share of cross-border cooperation and joint R&D projects within the automotive industry of the Greater Region is still relatively low.

The main barrier for cross-border activities is not, as it might be expected, the existence of different languages spoken within different parts of the Greater Region SaarLorLux. Rather, the enterprises perceive the lack of efficient support infrastructures as the single most important factor hampering the establishment of relationships across borders. The level of functional proximity is too low and therefore transactions costs to search potential partners and gather information about them are too high (Dahlman 1979), which inhibits intensive cross-border cooperation. Consequently, there is a strong need for platforms and organizations, which provide information and facilitate contacts between actors on different sides of the borders. The chapter also described numerous small activities mentioned by the interviewees, which seem to be capable of fulfilling such a broker function, and help reduce the distance between the actors. However, as most of these strategies are not new (Coenen, Moodysson and Asheim 2004, Trippl 2010) one question remains. Why, until today, has the cross-border functional proximity not been optimized, while the transition process of the automotive industry and a significant share of enterprises ask for it? This chapter has provided some answers to this question. The organizational structure of the sector, which is characterized by a large number of subsidiaries, hampers initiatives toward higher functional proximity. Furthermore, companies which are embedded in a supra-regional context are less dependent on an optimized proximity within the Greater Region SaarLorLux and do not invest all their energy in such initiatives.

In order to answer the question comprehensively and to understand why the prospects of cross-border cooperation within the automotive industry of the Greater Region have not yet been capitalized on intensively, a multi-level governance perspective and an evolutionary approach need to be adopted. As the statement of the cluster initiative's contact person shows, the initiative itself might be interested in low levels of proximity between automotive enterprises and companies from other industries or sub-regions, as this helps to keep competition at a minimum. Likewise, the role of policies at the regional, national and European level as well as their development over time has to be analyzed. What influence will the implementation of debt limits in Germany have on the room for maneuver of regional policies to facilitate cross-border contacts? To what extent are EU initiatives and funding efficient alternatives? How do such programs deal with conflicting interests among regional actors within the Greater Region SaarLorLux?

In a broader context, it will be interesting to study, and draw comparisons to, cross-border cooperation among automotive firms in regions characterized by more headquarters or R&D departments of OEMs and first tier suppliers. Furthermore, questions focused on whether the relevance of cross-border activities changes during different stages of the transition process and on the life cycle of the resulting new products deserve attention.

**References**

Andersson, T., Schwaag Serger, S., Sörvik, J. and Wise Hansson, E. 2004. *The Cluster Policies Whitebook*. Malmö: International Organisation for Knowledge Economy and Enterprise Development.

Automobilproduktion 2010. *Top 100 Automotive Suppliers – das Neue Ranking 2010/2011*. [Online]. Available at: http://www.automobil-produktion.de/top-100-2010-11/ [accessed: 11 November 2011].

Aydalot, P. 1986. *Milieux innovateurs en Europe*. Paris: GREMI [Online]. Available at: http://www.unine.ch/irer/gremi/Gremi%201.pdf [accessed: 11 November 2011].

Bathelt, H. 2006. Geographies of production: growth regimes in spatial perspective 3. Toward a relational view of economic action and policy. *Progress in Human Geography*, 30(2), 223–36.

Bilbao-Ubillos, J. 2010. Spatial implications of new dynamics in production organisation: the case of the automotive industry in the Basque country. *Urban Studies*, 47(5), 1117–46.

Boschma, R. and Frenken, K. 2009. The spatial evolution of innovation networks: a proximity perspective. *Papers in Evolutionary Economic Geography* [Online], 09(05), 1–16. Available at: http://econ.geo.uu.nl/peeg/peeg0905.pdf [accessed: 11 November 2011].

Coenen, L., Moodysson, J. and Asheim, B.T. 2004. Nodes, networks and proximities: on the knowledge dynamics of the Medicon Valley Biotech Cluster. *European Planning Studies*, 12(7), 10031–8.

Cooke, P. 2001. Regional innovation systems, clusters and the knowledge economy. *Industrial and Corporate Change*, 10(4), 945–74.

Dahlman, C.L. 1979. The Problem of Externality, *Journal of Law and Economics*, 22(1), 141–62.

Dörrenbächer, H.P. 2010. La «Gran Región». Institucionalización de una región europea tranfronteriza. *Documents d'Anàlisi Geogràfica*, 56(1), 185–200.

Dörrenbächer, P. and Schulz, C. 2005. Economic integration in the Saar-Lorraine border region, in *Borders and Economic Behaviour in Europe. A Geographical Approach*, edited by G. van Vilsteren and E. Wever. Assen: Royal Van Gorcum, 10–24.

Håkansson, H. and Johanson, J. 1993. The network as a governance structure. Interfirm cooperation beyond markets and hierarchies, in *The Embedded Firm*, edited by G. Grabher. London and New York: Routledge, 35–51.

Jürgens, U. and Krzywdzinski, M. 2009. Changing east-west division of labour in the European automotive industry. *European Urban and Regional Studies*, 16(1), 27–42.

Lundquist, K.-J. and Trippl, M. 2009. Towards cross-border innovation spaces: A theoretical analysis and empiricial comparison of the Orseund region and the Centrope area. [Online: IAREG Working Paper, 4.1]. Available at http://www. iareg.org/fileadmin/iareg/media/papers/IAREG_WP4.1_final.pdf [accessed: 11 November 2011].

Malmberg, A. and Maskell, P. 1997. Towards an explanation of regional specialization and industry agglomeration. *European Planning Studies*, 5(1), 25–41.

Markusen, A. 1996. Sticky places in slippery space: a typology of industrial districts. *Economic Geography*, 72(3), 293–313.

Martin, R. and Sunley, P. 2003. Deconstructing clusters: chaotic concept or policy panacea? *Journal of Economic Geography*, 3(1), 5–35.

Maskell, P. and Malmberg, A. 2007. Myopia, knowledge development and cluster evolution. *Journal of Economic Geography*, 7(5), 603–18.

Maxton, G.P. and Wormald, J. 2004. *Time for a Model Change – Re-engineering the Global Automotive Industry*. Cambridge: Cambridge University Press.

Morgan, K. 1997. The learning region: institutions, innovation and regional renewal. *Regional Studies*, 31(5), 491–503.

Perkmann, M. 2003. Cross-border regions in Europe: significance and drivers of regional cross-border co-operation. *European Urban and Regional Studies*, 10(2), 153–71.

Rutherford, T and Holmes, J. 2008. Engineering networks: university—industry networks in Southern Ontario automotive industry clusters. *Cambridge Journal of Regions, Economy and Society*, 1(2), 247–64.

Staatskanzlei Saarland. 2003. *Zukunftsbild 2020.* Saarbrücken: Chef der Staatskanzlei Saarland.

Sturgeon, T., Van Biesebroeck, J. and Gereffi, G. 2008. Value chains, networks and clusters: reframing the global automotive industry. *Journal of Economic Geography*, 8(3), 1–25.

Tödtling, F. and Trippl, M. 2004. Like phoenix from the ashes? The renewal of clusters on old industrial areas. *Urban Studies*, 41(5–6), 1175–95.

Torre, A. and Rallet, A. 2005. Proximity and localization. *Regional Studies*, 39(1), 47–59.

Trippl, M. 2010. Developing cross-border regional innovation systems: key factors and challenges. *Tijdschrift voor Economische en Sociale Geografie*, 101(2), 150–60.

VDA – Verband der Automobilindustrie. 2004. *Future Automotive Industry Structure (FAST) 2015 – die Neue Arbeitsteilung in der Automobilindustrie.* Frankfurt/Main: Verband der Automobilindustrie.

Chapter 4

# Industrial Restructuring by Financial Investors? Buy-and-Build Strategies in the German Automotive Supplier Industry from 2000 to 2010

Christoph Scheuplein

## Introduction

Private equity as a way of providing finance is seen by many as a key factor in the transition in today's relationships between the industrial and financial sectors. Since private equity collects global capital and fixes it locally for a while, it accelerates the transition of today's global industrial production structures. This applies in particular to Germany, where clearly observable changes have been taking place since the mid-1990s. These have included the decline in importance of the traditional 'house bank' system, the dissolution of inter-corporate links and the development of a new market for management control (Beyer and Höpner 2003). Private equity has had a significant role to play in these developments: the fund volume of domestic capital investment companies increased tenfold between 1992 and 2008 to a figure of around 35 billion € (BVK 2011: 26). It has been established that this involvement has had significant effects on macroeconomic performance, the majority of which have been positive (Kaserer et al. 2007: 49–53). For this reason there have been frequent calls for an improvement in the conditions for private equity in order to reinforce these positive effects. In short, the general view is that the private equity investment form has a great capacity for changing economic structures. In the public debate, private equity is now being viewed as the principal factor in the restructuring of 'Germany plc.'. Both proponents and critics of a more widespread use of private equity seem to be assuming that, in most cases, the logic of financial capital can dominate industrial capital.

The effect of private equity has in recent times been widely investigated in the empirical literature. These enquiries concern both the performance of private equity companies, along with the private equity funds managed by them, and the portfolio companies absorbed in the process, in whose case particular interest is generated by the situation of the employees (cf. as overview Cumming, Siegel and Wright 2007, Watt 2008, Strömberg 2009). Most empirical studies, however, present their arguments at the macroeconomic level or focus their attention on key values concerning profit, turnover or number of employees for a sample of private equity-led companies (for a critical appraisal cf. Vitols 2009: 71–78). A frequent problem with this approach is that it shows little understanding of the starting positions of individual enterprises: a blind eye is turned in particular to the sector-specific conditions of value creation. Yet, until these are taken into account, no judgment can be made as to whether the private equity enterprises are, along with a pure and simple optimisation of financial structures, also implementing restructuring strategies to improve the value added of the portfolio enterprises.

For this reason, the intention here is to adopt another research perspective, concentrating on the interfaces between the financial and industrial sectors in the framework of an investment process in a particular sector. This chapter is therefore concerned with one clearly demarcated sector and asks questions concerning the origin of the ability to restructure enterprises and the concrete effects that arise in the wake of such restructuring. The particular object of enquiry is the involvement since 2000 of private equity in the German automotive supplier industry. This is a large sector in Germany, which leads the way in both technological and economic terms and whose structure, based on medium-sized enterprises, makes it representative of many others. Recent years have in addition witnessed a significantly high number of takeovers. The period of scrutiny starts at the beginning of the decade, which can be identified as the starting point in the wave of takeovers in the sector. Also considered are the global economic crisis of 2008–09 and the ensuing period, as these represented a rigorous test of the portfolio enterprises and the mettle of their owners. It is above all the seismic shifts in the car industry that took place during this period which render it possible to make selective statements on the effects of private equity on an industrial sector.

This chapter adopts as its starting point the hypothesis that the effect of private equity depends very largely on specific sectoral structures and that conflicts may arise between the structures of private equity and the organisational and corporate structures of an industrial sector. It will be shown that private equity has committed itself to a long-term change in value added structures and in the process made extensive use of buy-and-build strategies. But this very same ambitious strategic approach has not resulted in success. Whereas the literature almost exclusively focuses on the strength of the financial sector, the relative tenacity of established industrial structures also needs to be considered.

The next section presents general assumptions on the effect of private equity in the industrial sector. Subsequently, these assumptions are considered in greater detail in the light of longer-term developments in the automotive supplier industry. This is followed by the presentation of a dataset concerning buyouts in the German automotive supplier industry, and the investment process that has run its course in the past decade is traced. Then the extent of the buy-and-build strategy in the automotive supplier sector is outlined, followed by an explanation of how this strategy foundered in the global economic crisis. The chapter concludes with considerations on the prospects for the future development of private equity in the automotive supplier sector.

## Value Added through Private Equity

The use of equity capital as an independent financing business model has in historical terms in Germany as elsewhere developed out of the financing of the set-up and growth phases of new enterprises (Jowett and Jowett 2011: 52–75). On the one hand, private equity now serves as a collective term embracing both venture capital and other related equity capital financing instruments outside the public capital market. On the other hand, a private equity buyout is understood exclusively in terms of the takeover of an established enterprise, and it is solely in these terms that the expression is used here. From the point of view of the purchaser, the reasons for a buyout of such companies vary greatly. These may include the search for a successor, the need to finance a corporate expansion or the unhappiness of the owners with the lack of success of an individual company. At issue may also be individual business units of a group which is looking to restructure its portfolio (Fenn, Liang and Prowse 1997: 27–34). If, as Joseph Schumpeter (1952: 153–52) asserts, credit serves to bridge the gap between products and the means of production, taking this as a basis makes it possible to distinguish between the following 'bridging functions' assumed by private equity in enterprises: These may be (a) a human resources bridge or (b) a financial bridge to cover a specific need (e.g. technological development, overseas expansion) or, in some cases, general liquidity requirements. A third option may be (c) an organisation bridge to another internal company structure or another business model. This bridge function hypothesis, which is used to formulate relevant tasks for the involvement of private equity at the microeconomic level, also implies a relationship between the financial and industrial sectors. For a limited time it makes sense for a type of financial actor to define and reorganise the structures of industrial capital.

It is possible here largely to fall into line with the business management literature, even though this uses the concept of levers of value creation to iron out the relevant distinction between value creation and value transfer. A corresponding  systematisation of the levers of value creation is frequently undertaken in accordance with management tasks such as financial engineering, operative business and strategic business (Berg and Gottschalg 2003: 17–25, Kaserer et al. 2007: 94–99, Vitols 2009: 62). Financial engineering is understood as an optimisation of the capital structure, which generally entails reducing tax payments and increasing the debt of the portfolio enterprise. Aspired to as operative tasks are generally a reduction in costs and an increase in revenues, the reduction of the capital requirement for operational business and the improvement of management skills. Those tasks frequently considered to be strategic include concentrating on core expertise and eliminating waste surrounding peripheral activities, but also boosting growth. Growth can in turn be pursued using a wide range of strategies, such as price or quality leadership. It goes without saying that a successful implementation of strategic approaches in particular can be expected also to have effects on the market and corporate structures of entire (sub-) sectors.

## Value Creation Structures in the Automotive Supplier Industry

In the large and venerable industrial sector that is the car industry, a complex structure of value creation has emerged. Of decisive importance is the value chain, which can be described in terms of material flow. It starts with the processing of raw materials, takes in the production of simple parts and components and culminates in the manufacture of vehicles by car manufacturers (Schonert 2008: 11–52). Characteristic of this chain is that parts and components are to a large extent created specifically for one model of vehicle, with the effect that product development, design, prototype construction and production planning have to be undertaken simultaneously for all the suppliers involved. This has led to the creation of a hierarchy, in which original equipment manufacturers (OEM) stand at the apex of the supplier pyramid without any limits to their influence. This preponderance of the OEM has increased significantly over recent decades through an intense process of concentration at the point of final manufacture. At the same time, the depth of OEM value creation has decreased, with the effect that the economic sphere of activity of the suppliers has significantly expanded. They have developed a greater depth of technological specialisation, and the number of independent enterprises has increased since the 1980s. Generally, three levels of supplier are distinguished according to the complexity of their products (Schonert 2008: 13). It is self-evident that this way of presenting the supplier pyramid will lead to a highly stylised picture. The number of actors does not undergo a linear change in accordance with the level, nor is it always possible to clearly categorise individual companies. At the end of the day, the enhanced skill of first tier suppliers in system integration in comparison with the OEMs has also made it more difficult to demarcate the former from the remaining categories.

There is a close correspondence between the car industry supplier pyramid outlined above and the logistical structure (Arnold 2008). Certain product classes are defined in terms of the frequency and standardisation of supply. Modules with the highest degree of specificity must be produced by suppliers within a radius of one or two driving hours of the OEM factory. This just-in-time logistical model has led to the establishment of supplier facilities in concentric rings around every volume manufacturer production site: with this arrangement, the organisation of production within the sector has also found a spatial expression.

In the supplier pyramids, the balance of power is asymmetrically weighted in favour of the OEM (Schonert 2008: 133–41). The reasons for this are to be found in the latter's access to end customers and capability of defining the brand or, as the case may be, individual model of car. The rigidity of the supplier pyramid for the suppliers manifests in turnover trends, participation in processes of innovation and the internationalisation strategy. Thus it is, for example, that the prospects for overseas expansion of many medium-sized suppliers are directly correlated with market entry on the part of an OEM.

OEMs now need, however, to make use of a sophisticated and interactive set of instruments to reinforce their positions of relative strength. These include communication on many operational levels and the exertion of influence using supplier manuals, certifications, financial and marketing incentives and quality controls. The constant growth in OEM product portfolios, shortening of product life cycle is ever higher expectations of vehicle properties have led in the car industry to the establishment between companies throughout the world of cooperative, simultaneous work processes. European OEMs have since the early 1990s increasingly been taking their cue from the cross-company networks of their far eastern competitors. They seek more than used to be the case to exert their influence using 'soft power'. Nonetheless, the suppliers continue to be subjected to constant price and quality pressures in the regular contractual negotiations and the OEMs are interested in ensuring that suppliers in all subsections of the supplier pyramid are locked into constant competition with one another. Market opportunities associated with a continuation of the outsourcing strategy and the assignment of module and systems competencies to the suppliers can only successfully be exploited through ever tighter integration into the supplier pyramid.

The wider framework conditions of the supplier sector also underscore the significance of this organisational structure. Firstly, stagnation in the car market in the triad triggers cut-throat competition for the available market share. By way of contrast, preferential access to the growing markets of the emerging economies, with particular reference to the BRIC states, can be gained via the OEMs. Secondly, there is a pressure to grow, derived, for instance, from increasing insistence on technological expertise and the ability to supply products worldwide. Thirdly, estimates of excess capacity at supplier level place it at one third of all capacity (Frick and Krug 2010:30). These factors combined have in the last decade led to high levels of takeover activity in the supply industry. In the wake of the recent recovery in the sector, it can be assumed that a new wave of mergers and acquisitions will follow. Due to the resultant trend to the creation of oligopolies in individual submarkets of the supply chain, this consolidation is being followed with intense interest by OEMs. In brief, along with the usual challenges of product innovation and business organisation, the portfolio companies of a financial investor in the car industry are also confronted with the reality that market access is only possible through their integration into the supply pyramids of individual OEMs. Boosting the value added of a portfolio company requires the improvement of its position in the supplier pyramid. This can be brought about by general business management methods such as work on innovative power, the efficiency of production or the internationalisation of the enterprise. But, for private equity companies in particular, this can be realised by the full exploitation of their core areas of expertise: the buying and selling of companies. This core competency is used in a growth-oriented and strategic manner in the buy-and-build strategy (BBS). This initially entails the purchase of a company to provide a platform, followed by further purchases of enterprises ('add-ons'), where such purchases should meaningfully boost or complement the technological or economic expertise of the first company. In the medium term, the enterprises should be integrated to form a new unit with a higher level of production capacity or product range, thus allowing synergistic gains to be made (Hoffmann 2008: 144–50). BBS is thus without doubt to be viewed as a plausible contribution to a sustainable increase in value added. At the same time, however, this strategy gives rise to obvious contradictions: Whilst the OEMs play an ambivalent role in the reorganisation of the value chain and the consolidation of the suppliers, a new actor in the form of private equity makes its entrance. An upgrading of portfolio companies in the supplier pyramid serves the medium-term financial aims of the private equity enterprise but may also run counter to the strategic aims of the OEMs. Prior to the discussion of how private equity-led takeovers have developed in this potentially explosive field, the next section will chart the past course of acquisitions by private equity enterprises of automotive suppliers in Germany.

## Private Equity Investments in the Automotive Supplier Industry

The German automotive supplier industry was at the turn of the new century a sector dominated by medium-sized enterprises that had grown in tandem with the OEMs, in which only a very few large companies were able to go head to head with the former. Whilst the activity of private equity at the macroeconomic level gradually increased, nothing more than piecemeal acquisitions took place in the automotive supplier industry. The year 2005, however, saw the publication of numerous magazine articles and sectoral studies reporting on a wave of takeovers. No single administrative or industrial association statistic can be cited as a basis for the investigation of this chapter. The German Private Equity and Venture Capital Association BVK (2011:18) for its part identifies the business and industrial products category under which the vast majority of automotive suppliers are classified. So, in order to arrive at an estimate of investments in the automotive supplier industry, one single database is used which is itself based on sector-specific information services, a special evaluation of data from the BVK, relevant company databases and other own research. Included were companies whose turnover is mainly generated by automotive sales, which employ in excess of 100 staff and which have their head offices in Germany. Also featured were subsidiaries of foreign companies which had gained a head office in Germany by virtue of the transaction. The decisive influence of the private equity enterprises was also in evidence. For this reason, silent participations and mezzanine financing were not classified as takeovers. Ownership of 25 per cent of the shares in a company was selected as the lower threshold. In most cases, however, the ownership share of the private equity enterprise was more than 50 per cent; in many cases they had complete control. Not taken into account were car producers, i.e. some manufacturers of special vehicles or sports cars, or producers of after-sales products. The period of evaluation of the database here runs from the start of 2000 to the end of the first quarter of 2011.

At the start of the decade, at least eleven companies were already owned by private equity enterprises. Since then, 163 corporate takeovers have taken place, with a peak in activity of 27 takeovers in 2005. As a raft of companies were sold several times over to a financial investor (secondary buyout), over the entire period 130 companies were at least for a time included in the portfolio of a private equity enterprise.

In these transactions, the companies that changed hands had a total turnover of some 24 billion € and combined workforce of roughly 165,000. If one includes the multiple transactions, the turnover volume increases to 30 billion € and the workforce to 210,000 employees. Measured by turnover and employee values, the years 2005 and 2006 can indeed be viewed as a 'wave' of takeovers. The two years when taken together account for 34 per cent of the employees and 36 per cent of the turnover of the entire period. After this boom, the financial crisis made a significant dent in the figures. In the years 2009 and 2010 only 22 transactions came to pass, and half of these were clearing-up operations: The companies had previously gone bankrupt with private equity enterprises as their owners. Investments in the automotive supplier industry after 2009 were characterised over and above the average of the private equity sector by post-insolvency restructuring and new starts.

If one is looking for a means of comparison of the 130 companies that came into the hands of financial investors between 2000 and 2010, the sectoral database 'Automobil-Zulieferer in Deutschland' ('Automotive suppliers in Germany') (Ehrig 2004) has much to recommend it. According to this source, in the first third of the decade around 1,050 companies with more than 100 employees were active on the market. It can therefore be assumed that roughly twelve per cent of the companies were acquired using private equity. These enterprises represented a cross-section of the whole supplier industry. Some three-quarters of all the companies had already been established prior to 1970; only in five per cent of the cases was the date of founding after reunification. Measured by turnover, the average among these companies was greater than that of all German automotive suppliers. The lowest segment (with a turnover from 10 to 24 million €) in particular is noticeably underrepresented in the portfolio companies. By way of contrast, roughly 46 per cent of the enterprises achieved a turnover in excess of 100 million € in the year of their takeover by a financial investor. This greater than average figure is not however replicated in the highest turnover class, with sales in excess of 1 billion €. This is an indication that acquisitions by private equity enterprises of first tier suppliers occurred with below-average frequency.

A similar picture is revealed in the workforce size categories. Here, too, smaller companies are underrepresented, and there is instead a definite emphasis on the third-largest size category (from 1,000 to 5,000 employees), which accounts for almost one-third of all enterprises taken over. In geographical terms, the suppliers which have been taken over are widely distributed, although there are discernable clusters in the typical supplier regions in Bavaria, Baden-Württemberg and North Rhine-Westphalia. Conversely, sites in eastern Germany barely feature at all. The product spectrum of the automotive supplier industry is completely covered by the portfolio enterprises; of the entire spectrum, the interiors (28 per cent) and exteriors (14 per cent) segments are most strongly represented. The engine, power train, chassis and bodywork areas each take a share of 10 to 12 per cent. Less favourably positioned, on the other hand, are the highly technical and fast growing fields of electrics/ electronics (7 per cent) and engineering services (1 per cent).

If the products and services are considered from the point of view of complexity, the picture that is revealed shows that, at the time of takeover, almost two-thirds of the enterprises were located on the third tier of the supplier pyramid. Roughly 30 per cent of the suppliers are to be viewed as 2nd tier suppliers and only 3.5 per cent can be allocated to the first level. It can therefore be seen overall that the majority of companies taken over in the automotive industry were larger, established enterprises, although it must be said that all sub-sectors of the industry were affected. It must also be acknowledged that the complexity of the products and the integration of the companies into the value chain in many cases have the effect of making the companies dependent in their supplier relations on the decisions made by companies above them in the hierarchy. The fact that enterprises with such a profile were selected can without doubt on the one hand be attributed to the purchase opportunities offered to the supply side. On the other hand, the favouring of the 3rd tier suppliers is also an expression of the buy-and-build strategy, which will be examined in greater detail in the following section.

**The Significance of Buy-and-Build in the Automotive Supplier Industry**

In the following, in each case of a merger of two automotive suppliers, the portfolio company which, from an economic and technological point of view, represents the more significant part of the merger is viewed as the platform enterprise. The less significant parts are designated as add-ons or development companies. Together these represent all the companies involved in a buy-and-build strategy. The first step is normally the acquisition of the platform company. However, as the later purchase of a more significant company can change the integration strategy, it is inadvisable for objective reasons to bind the concept too tightly to the chronology of events. A total of 55 companies or, as the case may be, 42 per cent of all portfolio companies can be designated build-and-buy enterprises. Measured by the number of employees and turnover, however, the proportion is slightly higher. Of these, 30 companies may be considered to be platform enterprises, whereas 25 served the purpose of development companies within this strategy. This ratio appears paradoxical only because the foreign companies added to a German platform enterprise are not listed as add-on companies and do not therefore form part of the investigation. In actual fact, the overall number of development companies is twice as high.

As the purchase of the platform companies usually precedes that of add-on enterprises, it becomes necessary to look beyond the nominal period of inquiry. A raft of companies, such as Empe-Werke Ernst Pelz & Co KG, Neumayer Tekfor Gruppe, JOST-Werke GmbH and Honsel AG, was acquired by financial investors in the second half of the 1990s, with the effect that, by the end of 1999, there was already an initial spread of portfolio companies. In 2000, five platform companies were added to the total, and, by the end of 2006, roughly 80 per cent of the companies which fulfilled the function of platform enterprises in the period of inquiry had been acquired. In retrospect, this acquisition of a series of large, technologically well placed suppliers in the first half of the decade can be considered an important reason for the sharp increase in takeover activity amongst SMEs in the middle of the decade. The years 2004 and 2008 did indeed witness the acquisition of more than two-thirds of all development companies.

Within the supplier pyramid, the build-and-buy companies can be distinguished from the other portfolio companies by virtue of a noticeably higher share of second-tier and a lower proportion of third-tier suppliers (Table 4.1). This difference is essentially down to the platform companies, of which over half are to be found on the second tier. Three-quarters of the add-on companies, on the other hand, are to be found on the third tier. Although the share of first tier companies amongst the platform companies is significantly greater than the average for the remaining enterprises, the ten per cent proportion of first tier companies, as a basis for a buy-and-build strategy, is not high. This shows that the apex of the supplier pyramid was not to any significant extent involved in a finance-driven restructuring process.

**Table 4.1**  **Ranking of portfolio companies of the automotive supplier industry in the supply pyramid in Germany in the years 2000 to 2010**

|  | Platform Companies (n=30) | Add-on Companies (n=25) | Total Platform and Add-on Companies (n=55) | Other Companies (n=75) |
|---|---|---|---|---|
| Tier-3 | 36,6 % | 76,0 % | 54,5 % | 70,6 % |
| Tier-2 | 53,3 % | 24,0 % | 40,0 % | 24,0 % |
| Tier-1 | 10,0 % | 0,0 % | 5,4 % | 5,3 % |
| Total | 100,0 % | 100,0 % | 100,0 % | 100,0 % |

*Source*: Own database

The product segments are widely dispersed in the buy-and-build companies, as they are in the overall sample. However, there is significantly greater focus on the interiors segment. One-third of all companies involved in a buy-and-build strategy can be assigned to this category. This tendency is particularly pronounced in the case of add-ons, a good half of which manufacture products for interiors. The bodywork area also carries greater weight in the buy-and-build companies. Alongside the exteriors field, which has slightly higher than average representation, buy-and-build strategies find application above all in those areas of industrial mass production with a moderate to low level of technological complexity. It can therefore be stated with confidence that private equity is of particular significance in sectors in which consolidation on the grounds of excess capacity is due. By way of contrast, companies from the electrics/electronics sector, with a share of the total buy-and-build field of only 2.4 per cent, are significantly underrepresented in comparison to the overall sample. The engineering services field does not feature: a fact which serves to underscore its failure as a technological pace-setter.

What can be expected is an extension of the holding period through buy-and-build – a state of affairs that reflects the complexity of the corporate management involved in a buy-and-build strategy. The average holding period per transaction for all buy-and-build companies is 43 months. If one excludes new acquisitions since the start of 2009, the holding period increases to 48 months. These two values are not very different from those of companies without buy-and-build. The values correspond to those established in other empirical investigations. In the sample taken by Becker (2009: 153), the holding time of those portfolio companies which had already been sold again at the time the survey was conducted was 47 months. An initial interpretation of the buy-and-build companies shows that there is no truth to the criticism of short-term business strategies often levelled private equity companies. Quick flips, i.e. the purchase and sales of a company in fewer than two years, hardly ever occur in either platform companies or add-ons, although they are by no means unheard of in the case of some of the remaining companies.

The holding time, however, varies sharply if one considers the biography of a company rather than individual transactions. Secondary buyouts prolong the holding time of the companies with buy-and-build to 65 months, whereas, in the remaining companies, the rise, to 53 months, is only slight. In this instance the platform companies are clearly of decisive importance, with an average holding time 80 months, twice as long as the development companies. Large platform companies such as Novem, Jost-Werke, Honsel und TMD Friction have remained in the hands of financial investors for periods of up to ten to eleven years.

If one considers the secondary buyouts that are responsible for these differences in holding time, the buy-and-build companies are subject to them significantly more often than the remainder. For instance, almost half of all buy-and-build companies were subject to a secondary buyout, whereas this was the case in only one-sixth of the remaining companies. Buy-and-build companies have also experienced tertiary buyouts considerably more often (8 per cent) than their non-buy-and-build equivalents (3 per cent). There are many reasons for this. The development companies are in part acquired from other financial investors. Once an initial optimisation of the acquired company has been successfully carried out, the buy-and-build strategy promises to generate further value. Furthermore, the platform companies, once acquired, are more readily available on the market for management control. In some instances, the opposite path is chosen: Once a company has been integrated into a new network, it is carved out anew and passed on to another financial investor who is confident of generating more income from a change in strategy. From time to time, the second financial investor continues along the buy-and-build track, so that the secondary buyout helps to meet the requirements of the long time frame of buy-and-build strategies.

An overview of all buyout activities in the German automotive supplier industry until the start of the global economic crisis of 2008 reveals that the buy-and-build strategy played a prominent role. It affected a good two-fifths of all companies and played itself out across the entire spectrum of product segments. The segments with the most ambitious technological developments (electronics and engineering services) were on the other hand underrepresented; most prominent were the manufacturers of parts for vehicle interiors. The build-and-buy strategies were aimed at a growth that was to be attained in the context of both completion of the product spectrum and horizontal integration. In some cases, growth was pursued exclusively by means of acquisitions overseas. At the same time, the intention was to use integration to improve the position of the companies in the supplier pyramid. These processes tended to take place in the middle and lower parts of the pyramid, whereby second-tier companies were preferred as platforms to which third-tier suppliers could be affiliated. Overall it can be said that, with this course of acquisition, the private equity companies played an active and strategic role whose ambition it was to change the value creation structures in the supplier industry. For this reason, better returns would have been yielded for the build-and-buy companies in comparison to the less complex strategies. This expectation was not fulfilled, although it must be said that the global economic crisis was a key conditioning factor in this outcome.

## The Failure of the Buy-and-Build Strategy

It took only a few weeks for the financial crisis that erupted across the world in September 2008 to have its impact on the real economy. The car industry was disproportionately affected in comparison to other industries, and most suppliers had to weather collapses in turnover of 40 per cent or more in 2009. Automotive suppliers began to file for bankruptcy in Germany as early as November 2008. The first heavyweight companies affected were Henniges Automotive, TMD Friction and Tedrive Germany GmbH (the German part of the former Ford supplier Visteon). There then followed further spectacular bankruptcies, such as those of Edscha, Plastal GmbH and the ac group in the first quarter of 2009. All the suppliers named above were owned by financial investors. And indeed, the rate of bankruptcy amongst companies owned by private equity enterprises was 10 times higher than that of companies with other owners. In this economic crisis in the automotive supplier industry, the large companies which filed for bankruptcy were almost exclusively owned by financial investors.

During the period of enquiry from 2000 to 2010, 43 of the 130 portfolio companies – almost exactly one third – went bankrupt. The buy-and-build companies fared no better than the others: on the contrary, with a bankruptcy rate of 44 per cent, they performed far worse than the remaining companies (24 per cent). The add-on companies, with a bankruptcy rate of 52 per cent, were more strongly affected than the platform companies (39 per cent). This shows that the platform companies which had made their acquisitions mainly overseas were more successful overall. In only 12 per cent of the cases was the large number of bankruptcies balanced out by the sale of a buy-and-build company to an industrial investor or by listing on the stock exchange. The buy-and-build companies therefore also fared worse than the remaining companies (19 per cent) (four of the five IPOs involved build-and-buy companies, but the total quantity of five IPOs during the entire period for all platform companies must have come as a severe disappointment).

The global financial crisis rendered an orderly exit between autumn 2008 and spring 2010 almost impossible. Accordingly, in the second quarter of 2011, half of all the companies were still in the hands of financial investors. The conditions created by the economic crisis also had an effect on the average holding time of the platform companies. The closing-off of the exit channel led to an extension of the holding time and can now be cited as an indicator of an unsuccessful exit strategy. As most funds have duration of seven years, the normal assumption is of a holding time of no more than five years in order to allow more time for the investment and disinvestment phases. The actual holding time of platform companies taken over between 2001 and 2003 has already risen to five years, and the holding time of the buy-and-build companies is even longer.

Overall it can be said that the buy-and-build strategy in the automotive supplier industry has met with little success. The situation in which the entire sector found itself can only bear limited blame for this, as, notwithstanding revenue difficulties, the supplier industry in the period of enquiry was able to record gains at least in turnover and manpower. If one is to look for the causes of this failure, the first port of call is the choice of companies. The majority of the companies had standardised production and moderate to low technological intensity. This means, for example, plastics injection moulding technology for interior and exterior vehicle parts, in which the development of technological USPs is problematic. In these areas, the private equity companies were aspiring to achieve a successful turnaround or consolidation in the individual product segments. In actual fact, however, the usual measures taken by the private equity companies in the car industry had less effect because the production and supply structures here are already closely coordinated at many levels. A purely cost-oriented acquisition strategy which involves a transition to global sourcing is limited by the fact that the OEMs already use their certifications and supplier manuals to restrict the choice of sub-suppliers. The production sites are frequently predetermined by the just-in-time logistics of the OEM factories, with the result that production relocations are only possible in individual cases (Wallentowitz, Freialdenhoven and Olschewski 2009: 59–73).

Conversely, the portfolio companies have to follow in the wake of the establishment of overseas production sites by the OEMs. The position becomes yet more difficult if suppliers are attempting to establish an autonomous technological position. The fact that the product life cycles of passenger cars are roughly 6 years means that there is an inevitable conflict with the shorter time horizons of the private equity companies. Furthermore, technological development in the car industry has long been driven by a complex process of cooperation which is organised by the OEMs. Here, too, a strategy of autonomy is confronted by numerous obstacles thrown up by the business environment. Private equity companies have however accepted these challenges, in particular with the buy-and-build strategy, setting and pursuing their own design objectives. What the 30 platform companies most clearly represented was an ambition to reorganise the supplier industry. In the process, some transactions were made on a very significant scale. However, hardly a single one of these platform companies successfully made the transition to industrial autonomy.

In the case of SAF-Holland GmbH, for example, it was possible in July 2007 to see through a partial exit; this was, however, only concluded in May 2010 with the complete withdrawal of financial investor. The intervening period witnessed not one successful exit, but ten platform companies filed for bankruptcy in the same period. It is striking in the context of the crisis hitting the sector that a large number of suppliers were shored up by OEMs and first tier suppliers. This took the form of an acceptance of higher prices, bridging loans and support in credit negotiations with the banks. Such aid was not, however, forthcoming to companies in the hands of private equity enterprises. A further example is the collapse of the Sellner group, which only took place many months after the renewed upturn in the automobile industry. The group was unable adequately to meet the growing demand, and quality problems and associated operative losses arose. Customers such as BMW, Audi and Daimler, along with the banks, chose in this instance not to participate in a bridging solution: instead they allowed the group to collapse. 'Finance', the sector's specialist periodical, has subsequently reported on the spreading of cheap propaganda by the OEMs and is of the opinion that financial investors have been subject to blacklisting (Schlumpberger 2011: 12).

In contrast, the OEMs take a thoroughly active role after the bankruptcy of suppliers and encourage industrial investors to make targeted takeover bids in order to safeguard the value chain. The large industrial corporations which participate in the consolidation also stress their coordination with the OEMs. These takeovers have ensured that most bankrupt companies have been able to continue to trade. At the start of the second quarter of 2011, more than half of the 55 buy-and-build enterprises were in the hands of financial investors and a little over a quarter had been bought up by an industrial owner. Only four companies – seven per cent – had been liquidated, and eleven per cent of the companies were still involved in an insolvency process. The switch to the public capital market or industrial ownership was successfully negotiated by 15 companies. Three of these companies are stock-market listed in free float and ten have successfully wooed foreign industrial investors, with China, with three investors, constituting the biggest single group. Further investors came from France, the USA, Canada, India and Austria. The return of the financial investor to industrial ownership has thus mostly gone hand-in-hand with the internationalisation of the owner.

## Conclusions

The biggest wave of takeovers in the German automotive supplier industry took place between five and seven years ago, a time frame which mirrors the normal lifetime of a private equity buyout fund. The acquisitions which took place at that time triggered in equal measure hopes of a thoroughgoing boost in efficiency and fears of a loss of self-determination and an uncontrolled drain of business assets. From today's perspective, it can firstly be seen that the financing structure of the acquired companies deteriorated because the credits raised to purchase the portfolio companies were imposed on the companies that had been acquired. This was, however, the main cause of bankruptcy in only a few cases.

A second finding is that, in the vast majority of cases, the portfolio companies were held by private equity companies for long periods of time. This is evidence that strategies were pursued for the long-term improvement of value creation in the portfolio companies. Buy-and-build strategies formed the backbone of these approaches, which were employed in the case of two-fifths of all companies acquired. The idea behind them was to improve the position of the portfolio companies in the supplier pyramid of the car industry. In doing so the private equity companies tried to bring about changes to the sector's value added structures in ways that would bring them benefit; the objectives of the acquisitions included higher market shares, greater system expertise or international market representation. Only in a few cases did this result in stable, newly integrated companies which could be floated on the stock market at a high mark-up or sold to industrial buyers. The result was instead that companies owned by financial investors went bankrupt much more frequently during the global economic crisis of 2008 to 2010 than companies with other owners. Worse still, companies in which a buy-and-build strategy was pursued were more adversely affected than the other portfolio companies. Some 45 per cent of all companies subjected to a buy-and-build strategy were in the meantime forced to undergo an insolvency procedure. The supplier landscape was as long ago as 2000 characterised by significant overcapacities, slender margins and a growing pressure on individual companies to expand and to internationalise their production. This resulted in significant pressure to consolidate the market, which by the end of the decade had led to new structures, with the greatest impact being felt in the first tier supplier field. It was in this of all segments, however, in comparison to the others, that the private equity companies were underrepresented. Nor did the five acquired tier 1 suppliers manage to drive the consolidation in any meaningful way: indeed, four of them have in the meantime largely been broken up.

The buy-and-build strategy has principally been implemented in second-tier suppliers to which further third-tier suppliers have been affiliated in order to allow the integrated companies to climb the ranks of the supplier pyramid. This led increasingly to agitation and turbulence in the power structures in the supplier pyramid. In the economic crisis, however, the balance of power shifted yet again. The OEMS used their market clout to rid themselves of the new actors. Whereas a financial safety net was strung under many of the automotive suppliers, such aid was not forthcoming for companies in the hands of financial investors.

With these preliminary conclusions in mind, what does the future hold for the involvement of financial investors in the automotive supplier industry? Private equity remains a significant player: half of all acquired companies are still in the hands of, or have been reacquired by, financial investors. At the same time, the increase in holding time would appear to indicate that, since the autumn of 2008, neither a buyer nor a satisfactory exit strategy has been found for a whole raft of portfolio companies. The fact, above all, that investments in the automotive supplier industry have resulted in losses has brought the business model of financial investors into question. The trend toward investment in this sector has slowed commensurately.

The significant diversity of developmental paths followed by automotive suppliers, with and without private equity owners, during the economic crisis underlines the relevance of the ownership function. At the same time, an assessment of the performance and results of financial investors in the context of sector-specific value creation structures reveals just why so many portfolio companies were so vulnerable to existential crisis. Whereas previous studies have used only general criteria to examine the effects of the investments from the point of view of private equity, the approach adopted here shows that the strategies pursued by private equity become problematic only against the backdrop of the supplier relationships, distribution of expertise and power relations characteristic of the particular structure of the industry. Whereas buy-and-build is normally considered the supreme discipline of private equity business, this strategy has under these conditions proved to be unworkable. The approach followed here instead posits that the limits of the business model have been harshly exposed by the tightly interwoven, systemically integrated production model of the German automotive supplier industry. In the context of the industrial transition, which is in the focus of this volume, this suggests independent paths on the national scale level. The accumulated global capital cannot be invested in all the places with the same measures, but it must be fitted into the national and regional production structure of an industry.

## References

Arnold, D. (ed.) 2008. *Handbuch Logistik*. 3rd Edition. Berlin, Heidelberg: Springer.

BVK 2011. *Das Jahr 2010 in Zahlen*. Berlin.

Becker, A. 2009. *Private Equity Buyout Fonds – Value Creation in Portfolio-unternehmen*. Bern: Haupt.

Berg, A. and Gottschalg, O. 2003. *Understanding Value Generation in Buyouts*. INSEAD Working Paper Series. Available at: https://studies2.hec.fr/jahia/

webdav/site/hec/shared/sites/buyoutresearch/acces_eleves/2003-42.pdf
[accessed: 28 October 2011]

Beyer, J. and Höpner, M. 2003. The Disintegration of Organized Capitalism: German Corporate Governance in the 1990s. *West European Politics*, 26(4), 179–98.

Cumming, D., Siegel, D.S. and Wright, M. 2007. Private equity, leveraged buyouts and governance. *Journal of Corporate Finance*, 13(4), 439–60.

Ehrig, K. (ed.) 2004. *Automobil-Zulieferer in Deutschland 2003/2004*. Landsberg.

Fenn, G.W., Liang, N. and Prowse, S. 1997. The Private Equity Market – An Overview. *Financial Markets, Institutions and Instruments*, 6(4), 1–105.

Frick, S. and Krug, V. 2010. Finanzierung, Konsolidierung, M&A und Strategie – Zulieferer in einem sich drastisch wandelnden Umfeld, in *Money vs. Technology. Wie die Finanz- und Absatzkrise sowie der technologische Wandel die Zulieferwirtschaft verändern wird*, edited by VDA. Berlin: VDA, 23–85.

Hoffmann, N. 2008. *German buyouts adopting a buy and build strategy: key characteristics, value creation and success factors*. Wiesbaden: Gabler.

Jowett, P. and Jowett, F. 2010. *Private Equity. The German Experience*. Houndsmills: Palgrave Macmillan.

Kaserer, C. et al. 2007. *Private Equity in Deutschland: Rahmenbedingungen, ökonomische Bedeutung und Handlungsempfehlungen*. Norderstedt: Books on Demand.

Martini, J. 2005. Keine Chancen für Familienbetriebe. *Automobil-Produktion*, 19(5), 32–36.

Schlumpberger, C. 2011. Überdreht. *Finance*, 12(4), 8–13.

Schonert, T. 2008. *Interorganisationale Wertschöpfungsnetzwerke in der deutschen Automobilindustrie: Die Ausgestaltung von Geschäftsbeziehungen am Beispiel internationaler Standortentscheidungen*. Wiesbaden: Gabler.

Schumpeter, J. 1952. *Theorie der wirtschaftlichen Entwicklung: Eine Untersuchung über Unternehmergewinn, Kapital, Kredit, Zins und Konjunkturzyklus*. 5th Edition. Berlin: Duncker and Humblot.

Strömberg, P. 2009. *The Economic and Social Impact of Private Equity in Europe: Summary of Research Findings*. Available at: http://ssrn.com/abstract=1429322 [accessed: 28 October 2011]

Vitols, S. 2009. Private equity: Financial engineering or solution to market failure, in *Finance-Led Capitalism? Macroeconomic Effects of Changes in the Financial Sector*, edited by E. Hein et al. Marburg: Metropolis. 2nd Edition, 61–87.

Wallentowitz, H., Freialdenhoven, A. and Olschewski, I. 2009. *Strategien in der Automobilindustrie: Technologietrends und Marktentwicklungen*. Wiesbaden: Vieweg und Teubner.

Watt, A. 2008. The impact of private equity on European companies and workers: key issues and a review of the evidence. *Industrial Relations Journal*, 39(6), 548–68.

Chapter 5

# Flexible Specialization – Thirty Years after the 'Second Industrial Divide': Lessons from the German Mechanical Engineering Industry in the Crisis 2008 to 2010

Martina Fuchs and Hanno Kempermann

## Introduction

Industrial transition is considered the economic, social and spatial re-organisation of industrial production and work against the background of growing global-local interdependencies. Not so long ago 'flexible specialization' was intensely debated as another new mode of production and work. At its heart, flexible specialization stands for 'post-Fordistic' production, implying innovative, knowledge-intensive and specialized production for one or a few clients, with close interaction and fine-tuning between manufacturer and client, creating economies of scope, going along with new kinds of labour relations and more qualified work and new strategies of outsourcing. Frequently, flexible specialization was proclaimed to replace 'Fordist' production, i.e. replaceable standard manufacturing, mass production and economies of scale.

Piore and Sabel (1984) in particular claimed 'The Second Industrial Divide' to be the new era after Fordism. The authors also regarded the new period as an opportunity for regional development, and they found historical examples of flexible specialized regions. They considered industrial districts, especially the garment industry of Northern Italy, to be an archetype of flexible specialization, but they also regarded the historical metalworking industry in the Bergisches Land in North Rhine-Westphalia and the medium-sized mechanical engineering companies in Baden-Württemberg as typical examples (Piore and Sabel 1984: 205–07, 216–20). After publication, the idea of flexible specialization played a central role in the sociological and geographical discussion on the limits of mass production, the end of 'Fordism' and the new age of flexible 'post-Fordist' production and regions. Nowadays, about thirty years after the discussion started, conceptual deficits as well as empirical trends put flexible specialization to the test.

The chapter will first show that the understanding of flexible specialization has changed since the 1980s at the conceptual level. We then argue that an up-to-date concept of flexible specialization must also take account of the fact that the institutional setting has changed during the last three decades.

Empirically, we focus on the crisis of 2008 to 2010 which served to sharpen the already harsh world market competition even more. The crisis can be seen as a hot spot in industrial transition because it challenges the organization of firms and threatens the strategies pursued by companies. The crisis accelerates processes and exacerbates recent trends, showing if – and which kind of – flexibility helps companies to survive.

## The Development of 'Flexible Specialization'

The concept of flexible specialization had its main starting point in the discussion about the regulation approach. The regulation school was designed to gain better understanding of the large recession in the world economy in the 1970s and 1980s (Aglietta 1976). The initial approach was soon extended to cover the more general topic of economic and social change, considering the limits of mass production, and the opportunities provided by flexible 'post-Fordist' production. In the 1980s and 1990s the concept was brought together with deeper analysis of economic flexibility by political deregulation in Europe and in the USA (Boyer 1987, 1995, Lipietz 1986). 'The Second Industrial Divide' (Piore and Sabel 1984) was in line with the economic-political discussion on a macro-level. In geography, mainly inspired by the regulation school and stimulated by insights with regard to the institutional pre-structuring of economic change, the discourse largely led to a critical politico-economic, institutionalist view (Jessop and Sum 2006). Even though Piore and Sabel (1984) examined the 'return of economics into society' mainly at a macro-level, the ensuing geographical discussion largely focused on the regional level, covering regional networks, industrial districts, or innovative milieus for example (e.g. Cooke 1997, Pyke and Sengenberger 1992, Sabel 1989, Saxenian 1992, Scott 1988, Storper and Bennett 1991). This theoretical path based on the political economic interpretation of flexible specialization explains 'post-Fordism' as the principal pattern of economic production, consumption and mode of institutional regulation, which continues to prevail in many core economies since the late 20th century. Although the political economic interpretation of flexible specialization as a result of the regulatory framework has remained highly topical, the institutional setting has changed. Analysis therefore has to take account of the new specific environment.

Strategic discussions are also held with regard to manufacturing and the shop floor. Some are closer to the neo-Schumperian tradition and the human relations perspective, others to critical labour sociology. Such divergent approaches are commonly linked by the insight that workers' participation, teamwork and various skills measures are increasingly essential assets in competition, because ongoing innovation activities are needed to meet customer demands and because workers share more responsibility for the production process. During the 1980s, the discussion surrounding 'post-Fordist' production became a practical as well as a theoretical one. In Germany, managers, works councils, trade unions and other actors accepted the normative premise that flexible specialization should evolve as a particular path of customer-oriented, permanent technological innovation related to new process organisation. Thus, various managerial strategies of flexible specialization were introduced, at least in some industrial sectors, companies or areas of production.

The following deliberations highlight three aspects, dealing with the results of such recommendations for flexible specialization and the ensuing strategies in the production process: First, 'flexibility' is not a general solution for companies, but a strategy leading to advantages as well as disadvantages. Second, the flexibility provided by the institutional environment of the company is an important issue. Today, the flexibility strategies and thus the working conditions for employees are influenced by new constellations with respect to both labour and capital markets. On the one hand, institutional deregulation in labour markets is providing more flexibility for companies, leading to a deeper segmentation between temporary workers and permanent employees; on the other hand, deregulation and increasing volatility of capital markets lead to restricted options for the companies and sometimes endanger jobs. Third, from a spatial point of view, the flexibility provided by the region is important with regard to labour, but less important when it comes to supply relations. This raises the question whether the local labour market is really an 'asset' for company flexibility, or seen from the labour perspective, more of a buffer zone with the local workers as an 'auxiliary army'. These three issues do not fundamentally call the use of the concept of flexible specialization into question, but they do need to be considered if the theoretical concept is to be brought up to date.

*Flexibility – Advantages and Disadvantages of the Strategy*

Piore and Sabel's (1984) assumptions on the advantages of flexible specialization clearly do not fit with later analysis. Surveys of the medium-sized mechanical engineering industry in Germany showed that flexible specialization, and thus the ability to deal with exceptionally high levels of complexity, can cost companies dearly. Knowledge-intensive single-part production for a particular client and the small-batch production for only some customers are particular cases of overly expensive 'over-engineering', meaning high-priced and costly over-specialization: Carrying out complex research, development and design for just a single client or a few customers prevents the company from achieving economies of scale (Herrigel 1997). Clearly, then, there are disadvantages in too high a degree of specialization, leading to counterproductive path dependencies and, possibly, situations of lock-in (Grabher 1993: 263–64, Hassink 2005).

As we will show in our empirical part, mechanical engineering companies attempt to solve the problem by developing and producing incremental innovations. They make use of their technological core competencies to adapt the product and thus broaden the market at least to some degree. Partly, this corresponds to 'flexible standardization' (Hirsch-Kreinsen 2009). Modularity plays a strong role. It essentially represents modular design of product and production process, designed to reduce complexity in production planning and control as well as obtain a continuous yet customer-oriented production process, connected to a 'deepening' of the markets. This shows the way to innovative paths of specialization (Glückler 2007), which is linked to a gradual diversification in the company's core competencies.

Last but not least, we have to be careful with our periodization of 'Fordism', 'post- Fordism' and 'industrial transition'. Flexible specialization is not completely new to the mechanical engineering industry in Germany. Many mechanical engineering companies still show a high degree of standardization today. Flexible specialization and the close cooperation of mechanical engineers and users in Germany dates back to the previous centuries. The engine building industry, which is considered the backbone of manufacturing, started to develop in Germany in the 1850s. Already at that time, many innovations were inspired by the clients and brought together different manufacturing sectors. Until the late 1920s, the market required no mass production, and was served by small and medium-sized firms. A certain kind of mass production developed due to the demand for complex mechanically automated systems in World War II. After the war, the German engine building industry began to grow again and reached its highest level of employment in the 1970s. Since then, the sector has had to economize and rationalize under the competitive pressures of globalization (Gretzinger 2008: 206). Thus, the path of innovation and flexible specialization began to be taken by German mechanical engineering companies well before the 20th century. On this broad road, the particular companies then follow their specific paths of flexible specialization.

*Flexibility Provided by the Institutional Setting: Labour Market Regulation*

In Germany, governments (with different party affiliations) launched various political instruments designed to increase the flexibility of employers in the labour market in the 1980s and 1990s. In that period, the institutional setting changed from public welfare towards incentives for private initiative. This specific institutional setting has encouraged company flexibility, especially in the crisis. In the labour market, however, it has led to considerable changes.

In the last decades, the standard employment relationship with often life-long work contracts and regular working hours continued to erode; at the same time the segmentation of the labour market into safe work contracts on the one hand and precarious jobs on the other deepened. Jobs of temporary staff and leased labourers became particularly precarious as they were no longer employed by the company itself, but increasingly subcontracted (or loaned) to the company by temporary work agencies. Short-time work is another important instrument for companies to keep their permanent staff. This is due to the specific German situation, which makes available state-funded short-time compensation for up to 24 months if production time has to be reduced for economic reasons or on account of another inevitable event (and if specific aspects of social legislation apply at the same time).

As we will see in the empirical analysis, apart from the institutional setting created by the government, it is also company-specific agreements on 'working hour accounts' that play an important role: working time accrued in the boom can be used up in times of recession, with 'accounts' sometimes becoming 'overdrawn'.

*Flexibility Restricted by the Financial Markets and Provided by the Liquidity and Capital Resources of the Company*

Boyer (1987, 1995) deals with financial relations as part of the regulation approach. In geography, Taylor and Thrift (1983) were among the first to point to the close connection between the financial sphere and the production process, although during the 1980s and 1990s neither social scientists nor managers could have anticipated the significance of financial markets and financial (de-)regulation for manufacturing industries in the recession of 2008–2010.

After deregulation of the financial markets in the USA and in Great Britain, which began in the 1970s and 1980s, new legislation was introduced in Germany in the 1990s and early 2000s that allowed broad use of new financial products. This resulted in increased opaqueness of transactions and risk potential. The ensuing speculation and chain reaction were the primary causes of the world economic recession. Financial (de-)regulation also shaped the effects of the crisis in the companies themselves. The new institutional setting made it easier for financial investors interested in short-term returns to take over companies. Independent manufacturing companies were weakened as a result.

In Europe, another factor that significantly changed the situation for small and medium sized companies was 'Basel II'. Introduced before the recession hit, this implemented regulations concerning the equity base of financial institutions. During the crisis, they now had an even stronger interest in maintaining their own equity base. As a result, credits for weak companies became expensive or were refused altogether. During the crisis, the credit stringency became a central problem for many small and medium sized companies in Germany, independent of the issue of flexible specialization in production. For them, flexibility was limited rather than enhanced by the institutional setting.

*Flexibility Provided by the Region*

In the spatial dimension, we take a critical view of the idea of flexible regional networks. Piore and Sabel (1984) and the debate that followed refer to industrial districts where local value chains exist between mostly small companies which manufacture products for common markets and often have mutually shared training centers and a specific creative milieu. Markusen (1996), however, highlighted the diversity of districts and their relations. Indeed, we know that the manufacturing industry mostly acts globally with regard to supply and sales markets. Even though we can find a few examples of local districts with internal, local value chains, they turn out to be rare in German manufacturing industries. As Herrigel (2002) stresses, most of the mechanical engineering companies in Baden-Württemberg had an extremely hard time on the world market during the 1990s, experiencing enormous pressure to increase their competitiveness. In consequence, there is no guaranteed insulation from the pressures of the world market, if this ever existed at all. Thus, neither in the 1980s nor today do Bergisches Land and Baden-Württemberg represent industrial districts in the sense of Piore and Sabel's typology (1984: 216–20). Yet in some regions such as Baden-Württemberg, Herrigel did find strong regional debates on the possibility and desirability of a future local economy; we will focus on such networks of regional actors and their discourse on regional specialization and diversification.

We therefore argue that Piore and Sabel's (1984) concept of 'flexible specialization' has to be revised to encompass a broader understanding of flexibility. Such flexibility mainly results from labour market policy, supplemented by in-house agreements. Furthermore, financial institutions influencing the liquidity and financial resources of the company are also important. Thus, the overall changes in the institutional environment on the macro, meso and micro levels need to be taken into account, as proclaimed by the political economic interpretation of the regulation approach. Last, but not least, the role of 'the region' has to be reconsidered: obviously, the relations of the value chains are often global, yet, human labour is essentially local.

## Methodology

Mechanical engineering is commonly understood to include manufacturers of durable means of production, or their pre-products and components, with the exception of electrical apparatus. We selected this sector for our examination because it is a key producer of investment goods. With regard to employment, the sector is by far the most important in Germany, employing about one million people (VDMA 2010). Mechanical engineering is also one of the strongest export sectors and rather research-intensive. As a result of the high innovativeness, the mechanical engineering industry has highly qualified employees. About 80 per cent of them are technicians and skilled engineers (VDMA 2009: 46–49).

We analyzed a dynamic situation during the crisis of 2008 to 2010. In our first survey, we calculated the coefficient of localization in mechanical engineering for different administrative districts ('Kreise'). The selected research regions with a high coefficient are primarily located in Baden-Württemberg and North Rhine-Westphalia. There, we selected two regions with a long tradition in the metalworking industry for qualitative research. These were regions we expected to be intensely impacted by the crisis on the basis of data and information collected in our first survey. We chose the districts Esslingen, Göppingen und Rems-Murr-

Kreis in the region of Stuttgart (Baden-Württemberg) and Siegen-Wittgenstein in the Bergisches Land (North Rhine-Westphalia). Thus, the 'post-Fordist' regions discussed by Piore and Sabel (1984) remain important locations for the mechanical engineering industry to this day.

As strategies of companies are a sensitive issue, our investigation had to be based on a qualitative approach. Qualitative methods are needed to discover the dimensions of flexibility; they cannot be exposed using standardised statistical methods. Also, the interactions of subjects and strategies were not foreseeable ex ante and thus had to be revealed through an explorative approach. Analysis and interpretation of the interviews was based on qualitative content analysis. We interviewed 20 company managers who agreed to an appointment. Furthermore, we interviewed 22 'regional' experts, such as works council members, representatives of the trade union IG Metall (the metal workers' union) and the employers' associations, agents of the engine building industry association and the chambers of industry and commerce, as well as experts employed in the economic development agencies. We were thus able to conduct interviews with the most prominent regional actors dealing with the crisis and the topic of flexibility.

## Empirical Insights: Flexible Specialization in the German Mechanical Engineering Industry during the Crisis

The following discussion illustrates that Piore and Sabel's idea of flexible specialization of production is still important, but needs to be modified to include the institutional setting (labour and financial capital) as well as a new understanding of the role of the region.

### Flexible Specialization of Production

The majority of the companies interviewed develop and design products according to consumer requirements, some as a single product and some in small-scale batches. The managers stated that close contact to the clients is important for their continued ability to innovate. Often, the close relationship to the client includes software design for the machine, as well as installation of the machine or after-sales service (maintenance, repair, updates, training). Servicing, in particular, contributes to strengthening customer relations. Interviews illustrated that companies able to cope with the crisis were successful customer-oriented producers: They either represented monopolists on the world market or world market leaders sharing an oligopoly with only a few other firms, sometimes with regionally divided markets. Pricing pressure is less strong than in the segment of standardized products. Some of the small and medium sized firms are 'hidden champions' on the world markets. Often, products are so complex that the client is more dependent on the producer than vice versa; flexible production combined with a superior position in the competition for innovation render them (nearly) irreplaceable to their clients. As the interviewed managers stated, sometimes clients are unable to cancel orders because there is no other provider and the clients lack the equipment or know-how to manufacture the machines themselves. Although this strong position helped the company to overcome the crisis, even successful companies have suffered a drop in incoming orders, leading them to introduce short-time work and sack 'loaned' temporary workers.

Many of the interviewed companies specialized in a core competency, e.g. pressure casting machines or milling. With a strong focus on a specific market niche, companies are hard pressed to realize economies of scale. Despite the advantages inherent in competing on innovation rather than price, in fact there is the danger of over-engineering, possibly combined with lock-in. Product modification is a relatively uncomplicated and economical way of developing new markets. Rather than broadening their markets, a deepening occurred, sometimes drawing in clients in completely new customer segments.

During the crisis, companies further developed and deepened their market-oriented strategies. About half of the interviewees mentioned different customer-oriented strategies. They spoke to their existing clients about new orders, financial opportunities, sales discounts, innovations which they might need etc. By doing so they hoped that in case of an upswing, their company would be top of the list for new orders. This strategy of contacting existing customers, however, encountered a problem when the client did not really need anything, or could not pay for a new order, and therefore did not respond positively to the request.

Apart from strengthening contacts to existing clients purchasing officers also tried to attract new clients. A producer of pressure casting machines for example, headquartered in the district of Rems-Murr, knew that three clients had gone public on the Chinese stock exchange in September 2008 and had thus become quite solvent. The German company contacted the Chinese firms, which led to new orders in March 2009, right in the middle of the recession. The purchasing officers also developed a successful new strategy for the firm and contacted clients all over Poland. Their argument was that clients would probably have more time for bargaining and testing new products in a crisis than in a boom.

The relevance of flexibility can be illustrated by contrasting it with standardized production. Standardized producers have limited elbow-room. The companies we interviewed often act in different market segments with different technologies, but none are technological market leaders with a unique selling point. Price competition is high; as a result there is high pressure on margins, resulting in a small equity base as well as limited means for R&D investment. Undersized financial cushions quickly erode in the crisis, and investments planned prior to the crisis are postponed.

Because innovation and product and process adaptations are important for survival, our interviews showed that most companies with flexible specialization continued their R&D much as before. Most did not introduce short-time work in their R&D departments. Rather than delaying ongoing R&D projects by reduced working hours they were completed speedily in order to bring the new products or product modification to the markets as quickly as possible. For these companies, keeping on their permanent staff represented a means of maintaining both their capacity to innovate and their customer orientation.

## Flexibility Provided by Labour Market Regulation

Though the mechanical engineering industry is a research-intensive sector with a large number of highly qualified, highly paid, workers, companies of the sector tend to spend less on manpower than on materials; on average, labour costs across the sector only amount to a quarter of all expenditure out of the gross production value.

However, in times of recession orders for materials can be cancelled much easier than labour contracts. From that point of view, temporary staff loaned by a temporary work agency is an adequate solution – at least in the short run since this staff usually do not have the qualifications, the 'tacit knowledge' and the experience of permanent staff. In Germany, there is also an advantage in terms of the company's figures because temporary staff is not counted as 'staff' in statistics. This means that each job converted from permanent staff to temporary lent staff leads to increases in productivity on paper.

In Germany, official statistics do not show the number of temporary workers by manufacturing sector. Generally, however, the number of temporary workers in the crisis fell by one third to one fourth. In the companies surveyed the share of temporary workers was between 10 per cent and 20 per cent; nearly all of them were made redundant and given back to the temp agency, which was usually unable to offer them alternatives. Employing staff on short term contracts (without a temp agency) is a rare occurrence in the companies surveyed because German law is very restrictive in this respect. In most cases short-term contracts are limited to apprentices. Nevertheless, most companies attempt to take on their apprentices after their apprenticeship because trainees are considered an important future resource.

Permanent staff was visibly less reduced in the mechanical engineering industry which is predominantly knowledge-intensive. 'Only' 4 per cent of permanent jobs were cut during the crisis (VDMA 2009, 2010). In the companies interviewed, only half the managers sporadically dismissed employees, especially those with too few qualifications and experience or maladapted behaviour. Only firms which were insolvent or nearly insolvent laid-off a higher number of workers. Even companies with a loss of turnover of 50 per cent and more laid-off staff as late as possible, if at all.

Short-time work is another instrument that allows companies to keep their permanent staff. In all sectors the number of short-time workers grew from 35,000 in the third quarter of 2008 to 1.5 million in May 2009; since then, the number has declined again (BA 2010). Mechanical engineering companies particularly make use of short-time work. A fifth of all short-time workers are employed in this sector, amounting to nearly 170,000 employees (BA 2010). Most of the companies interviewed introduced short-time work. Interviewees agreed that keeping permanent staff was important for the time after the crisis, when qualified engineers and workers would be needed to cope with new orders.

Another important strategy was to reduce permanent staff 'accounts' for working hours and to go 'overdrawn' wherever company agreements allowed. In the economic upswing before the recession, especially in 2007 and the first half of 2008, employees had worked on weekends and in the evening, sometimes doing shift work; these working time savings were spent now. Also, some companies stopped production in the crisis for a month, so all workers took annual leave in the same month. Therefore, at the company level there were agreements between employers and employees about working and production time to overcome the crisis.

### Flexibility Restricted by the Financial Markets and Provided by the Liquidity and Capital Resources of the Company

Innovation, playing an important role in flexible specialization, implies large investments. Therefore, the liquidity and capital resources of a company play a critical role. An executive of a world market leader with more than 5,000 employees illustrated this as follows: 'German mechanical engineering companies are often so strong in terms of innovation that the single long-term risk for dropping out of the market is their financial situation.' And a manager of another large company: 'Our company cannot be crowded out of the market just like that [because of the strong market position]; it can only become illiquid.' [Translation: M. Fuchs].

The equity base thus became an important factor in the crisis. Interviewees working in firms with an equity base of more than 50 per cent stated that they were not dependent on bank loans, and able to take independent decisions as far as investment, organisation and personal strategies were concerned. Such companies did not introduce short-time work in their R&D departments. Rather, their strategy was to keep being highly innovative throughout the crisis and to reap the competitive advantages afterwards. Some companies with a large equity base were able to make use of lower prices during the crisis and bought specialized machines and further equipment for very attractive prices. The situation was different in companies with standardized production. Already struggling with fierce price competition, they had only been able to build up a small equity base before the crisis.

Structures of decision-making and control in the companies are important factors, too. Many of the small and medium sized firms as well as some large mechanical engineering companies are family-owned, but many companies whose names suggest otherwise are actually part of a national or international trust, or division of an investment firm. In general, the German mechanical engineering industry does not consist of small and medium sized firms – on the contrary: within the manufacturing industry, the mechanical engineering sector comprises more large companies than all the other sectors. In consequence, some companies had to implement investment decisions taken in 'external' headquarters, decisions they could not afford in the crisis. Other companies are part of private equity firms with a strong focus on shareholder value, which also leads to small financial cushions. Earnings made in good economic times could not be used for building up a financial cushion.

Apart from being dependent on headquarters, mechanical engineering companies also suffer from dependency on the client. This is particularly the case with companies supplying the automobile industry. As it is customers that dictate investments, prices, batches etc., there was little scope for these companies to increase their financial cushion before the crisis. Capacity utilization is also highly dependent on the automobile firms. When the automobile companies changed their plans, their providers had to follow.

However, independence in decision-making – with no influence from headquarters, clients, or banks – is no guarantee for steering out of the crisis. For some companies for example, the good economic times before the crisis were too short to build up a solid financial cushion. The most important problem, however, was that some managers did not realize the dimension of the crisis early enough. Convinced that incoming orders would increase as they had in 2006–2008, they invested in capacities, bought pre-products for their stocks, and forgot the long-term limits of growth. Since many managers shared this optimistic perspective on growth, interviewees were in line with the overall opinion in the sector, leading to risky decisions.

*Flexibility Provided by the Region*

Regional value chains, stressed strongly in the discussion following Piore and Sabel (1984), only play a marginal role. Most of the companies interviewed buy their pre-products and sell their products nationally or internationally; regional connections are incidental. There are no clusters in the strong sense, i.e. with strong internal complexity. Thus, flexibility provided by local value chains is rare.

Still, some managers developed regional strategies for increasing the level of qualification of their employees and supporting young professionals. We found this in different regions of our empirical study in Baden-Württemberg and North Rhine-Westphalia. At the local level, contacts were cultivated in meetings, with company managers and works councils, with the employers association and the trade union, the employment office, in some cases also with chambers of commerce and industry and the mechanical engineering association. In some cases research institutes dealing with mechanical engineering were involved, while in other cases research and consulting institutes dealing with socio-economic regional change played an active role.

For example, some companies joined forces with the employment office to develop a job creation company, where young and unemployed engineers and apprentices could work and learn. The job creation company does not only benefit young people who can learn and stay in work, but also the companies which want (and have) to continue with their innovation and R&D to improve their flexibility. Thus, the region is clearly more than just a buffer stock of the labour force, because in some cases the actors seek to develop solutions and win-win situations with other actors to hold on to the workforce and to further improve their qualifications. In the Stuttgart region in particular the interviewees did not view the local crisis of 2008–2010 to be the result of the overall recession. They saw it as an effect of regional specialization in the automotive industries, where companies had become locked into too high a degree of specialization and excluded from flexibility through diversity. As a member of the chamber of industry and commerce put it: 'This (…) is not mainly a crisis of the mechanical engineering industries, it is a crisis of our manufacturing industry' [Translation: M. Fuchs]. A member of an economic development agency stated: 'The district of Göppingen is not in an economic recession, but in a structural crisis. We have known for more than thirty years that we are excessively dependent on the automobile industries. This means that many of our companies, nearly all companies, are related to the automotive value chain. They are suppliers of specific parts, or – on a higher level of the value chain – of machine tools. 80 to 90 per cent of the machine tools we produce are sold to the automobile sector.' [Translation: M. Fuchs].

Such common understanding of the crisis as a problematic path, or lock-in resulting from regional specialization, is a product of discourses that have taken place in the Stuttgart region since the 1980s. As mentioned by Herrigel (2002), diverse regional conferences, meetings, workshops etc. dealing with regional change have brought the different actors together during the last thirty years. The various reunions helped to at least partially overcome conflicts of interest between employers and employees and their representatives in a pragmatic way. Some interviewees used the expression 'Spaetzle connection', which refers to a typical local spaghetti speciality. The trade union (IG Metall), works councils and a regional research and consulting organization, in particular, developed ideas on how to diversify the region, especially towards e-mobility as a new regional innovation path, trying to combine improvements in the labour market with innovations said to be sustainable with regard to ecology and climate change.

In the other selected region, Siegen-Wittgenstein, the situation was quite different. In this traditional metal-working region, comprising a large number of small cities and rural landscapes, managers, works councils, trade unionists etc. cooperate, too: 'Practically, we live in a village. Everyone knows everyone, or somebody knows something', as a local manager told us. Unlike the Stuttgart region, however, the local discourse did not refer to a problematic path of regional specialization. In Siegen-Wittgenstein the prominent actors (managers, works councils, the trade union IG Metall, employers' association, chambers and associations) witnessed an early end of the crisis. An important meeting was a conference of works councils and IG Metall, which found that only few employees were affected by crisis and that the economy was generally picking up in the region, although some companies were hit by the crisis. The local actors agreed that the recession of 2008–2010 was not important and returned to their previous tasks. This response is related to the regional companies' specialization on large plant engineering and construction. Such companies usually have large orders that allowed them to continue working through the crisis; later they acquired new orders. Thus, perception by regional actors also plays a role in flexibility, in that they will only act if they see a need to adapt to changing conditions. Such awareness, in turn, is related to the discourse in the regional network and how this discourse develops over time.

## Conclusion

Even if the outline of the industrial transition is still somewhat fuzzy, we perceive the first vague contours of a more complex concept of flexibility in the 'third industrial divide', which could not be elaborated in this qualitative study in detail. The fierce competition in the global markets and especially in the economic recession put the strategy of flexible specialization to the test. The chapter discussed to what extent flexible specialization still plays a role today, especially in times of recession. We found that 'flexibility' is not a general solution for companies, but a strategy leading to advantages as well as to disadvantages.

The flexibility provided by the institutional environment is important issue. On a macro-level institutional deregulation providing more flexibility to labour markets is central. In consequence, local labour markets are (further) divided into permanent and temporary staff, with growing gaps between the segments. The deregulation of the financial markets restricted the elbow-room for the companies and led to greater importance for the liquidity and capital resources of the companies. In providing flexibility, the equity base of the company becomes a significant institution at the micro-level. Some companies with a large equity base were even able to profit from the crisis, e.g. by making use of lower prices during the crisis. However, most of the mechanical engineering companies hit by the recession were part of private equity firms with a strong focus on shareholder value.

From a spatial point of view and on the mesolevel, the flexibility provided by the region is important, especially with regard to labour and qualifications. While regional value chains rarely play a role in our case study, regional cooperation is a key to preserving employment as well as training and apprenticeships. Contacts were cultivated in meetings between managers and a range of regional actors that play a role in the local labour market. Sometimes, as in the Stuttgart region, the common discourse has led to joint regional strategies to overcome lock-in and to promote regional innovation in new market segments.

Thus, we find that flexibility provided by the region is only a specific spatial 'layer', and does not necessarily imply that the region is used as buffer stock for labour. As qualified labour is important, local actors search for common strategies to keep and improve the workforce.

In sum, what seems to be particular with regard to the industrial transition is the insight that the concept of flexible specialization seen from the perspective of political economy still matters 'thirty years after'; yet, it has to be refreshed by integrating new institutional settings and by reconsidering what is meant by 'the region'. Besides the mesolevel, companies attain flexibility by combining different scales; the national labour market and the 'in-house' flexibility of capital and liquidity resources play significant roles, too.

## References

Aglietta, M. 1976. *Régulation et crises du capitalisme. L'expérience des Etats Unis*. Paris: Calmann-Lévy.

Baca, G. 2004. Legends of Fordism. Between Myth, History, and Foregone Conclusions. *Social Analysis*, 48(3), 169–78.

Boyer, R. 1987. *La théorie de la régulation. Une analyse critique*. Paris: Découverte.

Boyer, R. 1995. Aux origenes de la théorie de la régulation, in *Théorie de la régulation. L' état des savoirs*, edited by R. Boyer and Y. Saillard. Paris: Découverte, 21–30.

BA 2010. *Arbeitsmarkt in Zahlen. [Labour market statistics]*. Nürnberg: BA.

Challet, D., Solomon, S. and Yaari, G. 2009: The Universal Shape of Economic Recession and Recovery after a Shock. *Economics*, 3(36), 1–24.

Cooke, P. 1997. Regions in a global market: the experiences of Wales and Baden Württemberg. *Review of International Political Economy*, 4(2), 349–81.

Grabher, G. 2009. Yet Another Turn? The Evolutionary Project in Economic Geography. *Economic Geography*, 85(2), 119–27.

Gretzinger, S. 2008: Strategic outsourcing in the German Engine Building Industry. An Empirical Study Based on the Resource Dependence Approach. *Management Revue* (19)3, 200–28.

Herrigel, G. 1997. The Limits of German Manufacturing Flexibility, in *The Political Economy of Unified Germany: Reform and Resurgence or Another Model in Decline?*, edited by L. Turner. Ithaca: ILR-Cornell University Press, 177–206.

Herrigel, G. 2002. Large Firms and Industrial Districts in Europe: De-regionalization, Re-Regionalization and the Transformation of Manufacturing Flexibility, in *Regions, Globalization and the Knowledge Based Economy*, edited by J. Dunning. Oxford: Oxford University Press, 286–302.

Hirsch-Kreinsen, H. 2009. *Innovative Arbeitspolitik im Maschinenbau? [Innovative labour market policy in engine building industry?]* Dortmund: University Press.

Jessop, B. and Sum, N.L. 2006. *Beyond the Regulation Approach. Putting Capitalist Economies in their Place*. Cheltenham: Edward Elgar.

Lipietz, A. 1986. *Mirages et miracles. Problèmes de l'industrialisation dans le tiers monde*. Paris: Découverte.

MacKinnon, D., Cumbers, A., Pike, A, Birch, K. and McMaster, R. 2009. Evolution in Economic Geography. Institutions, Political Economy, and Adaptation. *Economic Geography*, 85(2), 129–50.

Markusen, M. 1996. Sticky Places in Slippery Space: a Typology of Industrial Districts. *Economic Geography*, 72(2), 294–314.

Martin, R., Sunley, P. 2010: The place of path dependence in an evolutionary perspective on the economic landscape, in *Handbook of Evolutionary Economic Geography,* edited by R. Boschma and R. Martin. Cheltenham, Northampton: Edward Elgar, 62–92.

Pike, A., Birch, K., Cumbers, A., MacKinnon, D. and McMaster, R. 2009. A Geographical Political Economy of Evolution in Economic Geography, *Economic Geography* 85(2), 175–82.

Piore, M.J. and Sabel, C.F. 1984. *The Second Industrial Divide. Possibilities for Prosperity*. New York: Basic Books.

Pyke, F. and Sengenberger, W. (eds.) 1992. *Industrial districts and local economic regeneration*. Geneva: International Institute for Labour Studies.

Sabel, C. 1989. Flexible specialization and the re-emergence of regional economies, in *Reversing industrial decline? Industrial structure and policy in Britain and her competitors,* edited by P. Hirst and J. Zeitlin. London: Routledge, 17–70.

Saxenian, A. 1992. Regional Advantage. Cambridge: Harvard University Press.

Scott, A.J. 1988. *New Industrial Spaces: Flexible Production Organization and Regional Development in North America and Western Europe*. London: Pion.

Storper, M. and Bennett, H. 1991. Flexibility, hierarchy and regional development: The changing structure of industrial production systems and their forms of governance in the 1990s. *Research Policy*, 20(5), 407–22.

Taylor, M. and Thrift, N. 1983. The Role of Finance in the Evolution and Functioning of Industrial Systems, in: *Spatial Analysis, Industry and the Industrial Environment. Vol. 3, Regional Economies and Industrial Systems*, edited by F.E.I. Hamilton and G.J.R. Linge. Chichester: Wiley, 359–85.

VDMA 2009. *Statistisches Handbuch für den Maschinenbau. [Statistical handbook of engine building industry.]* Frankfurt/Main: VDMA.

VDMA 2010: *Maschinenbau. [Engine building industry].* Frankfurt/Main: VDMA.

Chapter 6

# Organisation and Representation of Temporary Workers

Dorit Meyer

## Introduction

The phenomenon of decreasing union membership, apparent in Western industrial countries since the 1980s, can be in particular explained by structural changes. Deindustrialisation has eroded the traditional union membership base, because the shrinking industrial sector used to constitute the core of union organisation. In addition, precarious and atypical forms of employment, such as temporary employment, are becoming more widespread (Coe et al. 2008, Peck and Theodore 2007). However, these types of workers still remain underrepresented by unions. Exploiting the new membership potential is one of the prime objectives for unions. This is the only means by which to halt their decline in membership and strengthen their bargaining power (Frege and Kelly 2004).

As one of the new target groups for union recruitment, temporary workers are beyond the reach of conventional recruitment routines. The tried and tested routines for promoting worker interests, such as the organisation of strikes, are not suitable for achieving advancements within temporary employment, because temporary workers have worries that they will be dismissed if they fail to deliver the demanded performance. Another reason why temporary employment presents a particular challenge for unions is that the economic crisis in 2008 and 2009 impacted temporary workers especially severely and led to mass lay-offs.

With regard to the organisation of atypical workers in general, there are still serious gaps in 'labour geography' research (Herod 2010: 24). But advocates of interdisciplinary research on union revitalisation (e.g. Frege and Kelly 2003) consider it thoroughly possible to achieve a reinvention of organisations with learning processes and new political practices. Taking the Industriegewerkschaft Metall (IG Metall) in Germany as an example, this chapter makes a contribution to research into trade unions' responses to temporary employment.

Unions have to learn the capability of organising temporary workers and representing their particular interests. To this end, new routines need to be developed which are tailored to the specifics of temporary employment. This chapter puts forward the argument that the dynamic capabilities (Teece et al. 1997, Eisenhardt and Martin 2000) of IG Metall have placed it in a position to do this.

Based on empirical results, I provide a response to the research inquiry as to how dynamic capabilities have enabled IG Metall to react to the increasing spread of temporary work prior to the start of the economic crisis. I also show how the adaptation responses to the changed conditions and new requirements during the crisis can be explained with dynamic capabilities.

First, this chapter will give an overview on the concept of dynamic capabilities and show under what conditions routines and organisational capabilities can be developed. Because organisations are structured on a multi-scalar and multi-location basis, different bottom-up and top-down learning processes occur in various locations and scales of the union organisation. In addition, I will delineate dynamic capabilities by definition from ad-hoc reactions. Second, based on the empirical data the multi-scalar as well as multi-locational resource developments with dynamic capabilities in bottom-up and top-down learning processes in the IG Metall will be outlined.

## Dynamic Capabilities

Prior to the recent crisis, the union responded to the increasing numbers of temporary workers in Germany by establishing new resource configurations. As the following sections explain, accomplishing resource development processes within organisations is only possible with dynamic capabilities.

### The Concept of Dynamic Capabilities

The resource-based view (e.g. Barney 1991) of organisational sciences emphasises that competitive organisations are each equipped with value-creating, rare, inimitable and organisation-specific resources (e.g. knowledge). Organisations use their resources in a particular way in order to create competitive advantages.

However, in the resource based view resources are regarded as unchanging and dynamics within the operational environment of the organisation are not considered (Helfat and Peteraf 2003). The resource based view merely describes how organisations achieve a superior competitive position at a particular time but does not explain how organisations are able to secure their competitive advantages in the long run when faced with changing conditions in their environment (Eisenhardt and Martin 2000: 1106). The organisation is ultimately required to develop its resources efficiently in order to prevent the erosion of its competitive advantages (Helfat et al. 2007: 1). It is because of this static approach that the resource based view has become the subject of criticism (Teece et al. 1997: 514).

The concept of dynamic capabilities taps into this debate and develops the resource based view further. This new approach also allows for an explanation of the competitiveness of organisations under changing conditions, because it takes into account the intention of the organisation to transform, to adapt, be flexible and innovate through organisational learning processes (Teece et al. 1997, Eisenhardt and Martin 2000). Dynamic capabilities put organisations in a position to establish innovative problem solution strategies and to improve efficiency (Zollo and Winter 2002: 340). Only with these can the organisation remain successful when the environmental conditions and demands change.

In order to exploit this option, the organisation uses its dynamic capabilities in particular to develop resources. The integration of new resources and the recombination of the resource base within learning processes create a new resource configuration that is applied to the changed conditions in the environment (Helfat et al. 2007: 5). The resources that are developed with dynamic capabilities consist of routines and organisational capabilities (Teece et al. 1997: 515, Eisenhardt and Martin 2000: 1107). The following sections discuss only these resources.

Routines are repetitive patterns of behaviour for solving recurring problems. As standardised sequences of activities that have a particular objective, routines help individuals to deal with typical tasks in an automatic way and thus serve the actors in the organisation as an orientation aid (Nelson and Winter 1982). In order to apply routines systematically and combine them efficiently, an organisation must utilize its organisational capabilities (Helfat and Peteraf 2003: 999). Capabilities also enable the organisation to coordinate activities and routines in a targeted manner, through which processes are performed within the operative day-to-day actions (Helfat et al. 2007: 1).

In order to adapt to changes, new resource configurations are formed from routines and organisational capabilities through organisational learning in resource development processes with dynamic capabilities. Already existing resource constellations are modified or reconfigured (Teece et al. 1997, Eisenhardt and Martin 2000).

## *Dynamic Capabilities in the Multi-Location and Multi-Scalar Union Structure*

Large organisations consist of individual sub-entities at various locations in the individual structure of the organisational administration (Wiesenthal 1995). Unions, too, are comprised of sub-entities (Cumbers et al. 2010). Consequently, the resource development processes with dynamic capabilities do not proceed uniformly, but rather in a multi-location and multi-scalar manner, as is described below.

'Multi-location' indicates that trade union secretaries participate at various union locations with various degrees of activity in the use of dynamic capabilities for the development of resources. The reason for this difference is that in larger organisations the entire organisational structure is not uniformly impacted by the same influences (Lévesque and Murray 2010: 237). Instead, the sub-entities of an organisation are confronted with different environments and are occupied to varying degrees with adapting, learning and innovating. Because the intensity of the commitment of local union entities differs with regard to the application of the dynamic capabilities, geographical 'patterns of union involvement' (Lévesque and Murray 2010: 225) become established.

The fact that resource development with the dynamic capabilities takes place on a 'multi-scalar' basis refers to the structure of the trade union organisation, which is comprised of three hierarchical scales (local administration offices, regional districts and the national federal board) (Herod 2010: 16). Attempts by unions to innovate receive the most support and are performed most effectively if they are implemented at various hierarchical scales (Wills 2005). To this end, the various scales have to contribute to the learning processes with their scale-specific knowledge and resources (Tattersall 2006: 14).

### *Prerequisites for Resource Development using Dynamic Capabilities*

As previously mentioned, the core remit of dynamic capabilities is the development of resources for adapting to changing environmental conditions (Teece et al. 1997). Routines and organisational capabilities are formed from knowledge in learning processes (Levitt and March 1988) and are therefore termed knowledge-based (Nelson and Winter 1982, Teece 2009). In order to enable resource development processes using dynamic capabilities, the following prerequisites must be fulfilled.

First, change must occur in the environment of the organisation. For example, union activity is always embedded in institution frameworks which frequently change (Teece 2009: 11).

Second, according to the findings of labour geography (Tattersall 2006) and research into dynamic capabilities (Teece 2009), particularly committed actors within the organisation must exhibit a willingness to initiate change. This also accords with statements made in publications about organisational learning: 'Organizational learning occurs when members of the organization act as learning agents for the organization, responding to changes in the internal and external environment' (Argyris and Schön 1978: 29). Such promoters (Teece 2009: 19) identify inefficiency in the use of existing routines and then use dynamic capabilities to develop new routines.

Third, the promoters must be embedded in networks external to the organisation for them to be able to access urgently required knowledge (Herod 2010: 23). Promoters cooperate with some of the members of the network in resource development.

Fourth, favourable opportunity structures are required: Within a rapidly changing environment, new opportunities for organisational change emerge frequently. Promoters recognise these opportunities by constantly searching for them within their operating environment and seize them by initiating learning processes (Teece 2009: 9).

Fifth, changes within the organisation can exert pressure, causing some actors to commence resource development processes, as is described in more recent publications concerning union-related change (Frege and Kelly 2003: 15).

Hence, not only external but also internal factors influence if and where which resources are developed. These prerequisites are not satisfied at every union location but vary from place to place (Herod 2010).

*Resource Development Bottom-Up*

With dynamic capabilities, the need for change is recognized and an appropriate reaction generated which demands the development of resources (Eisenhardt and Martin 2000). If, following the application of a routine and use of an organisational capability, there is a gulf between the expected and the actual result of the action this becomes the trigger for organisational learning (Argyris and Schön 1978). Existing resource configurations will be critically reflected upon and the development of resources will be commenced within the learning processes. In organisational research literature there is mention of various learning processes that are necessary for the dynamic capabilities.

In learning-by-searching, a promoter systematically searches for relevant new knowledge which is of strategic importance for the organisation and which ultimately could lead to an improvement of existing processes. In learning-by-interacting (Gertler 2003), new knowledge is acquired through communication and/or cooperation between actors. Their spatial proximity leads to frequent face-to-face contact and thus an exchange of knowledge. If the actors realise they are committed to the same objective, then the interaction will frequently consist not merely of the communication-based exchange of knowledge, but also a joint commitment towards a particular issue. In the case of experimental learning (Eisenhardt and Martin 2000: 1107) the actor develops reactions within the organisation and tests their effectiveness in new situations. If the innovative approach leads to success it is repeated and established as a new routine. In learning-by-doing (Levitt and March 1988: 321), not only intended learning effects and experiences result, but also other coincidental ones as 'by-product' of day-to-day activities.

In these learning processes new knowledge is integrated into the organisation. It is then used to build new knowledge-based resources and to further develop existing knowledge-based resources. The resource configurations in the organisation are thereby changed. However, for this purpose the individual results of learning must first be supplemented by collective learning processes. To this end promoters must be integrated in networks internal to the organisation on the same and on other hierarchical scales. In these networks, promoters introduce the new knowledge into internal discussions of the organisation (Zollo and Winter 2002), e.g. at the union's annual congresses.

*Resource Development Top-Down*

The application of the dynamic capabilities is not completed with individual and collective learning efforts. The new knowledge-based resource configurations must first be disseminated within the organisation with the board specifying in written guidelines what routines are to be used to surmount problems arising within the dynamic environment (Zollo and Winter 2002: 342). Because the leadership of an organisation promote innovations, motivate employees, develop control mechanisms and fundamentally coordinate the operations, they are attributed a high significance in the application of dynamic capabilities to disseminate new resources (Eisenhardt and Martin 2000: 1107).

*Dynamic Capabilities in Crises*

Crises occur rapidly and result in new challenges to organisations (Eisenhardt and Martin 2000). It is disputed, however, whether the reactions of organisations to crises can be explained with dynamic capabilities. A dynamic capability is a complex, stable model of systematic resource development (Teece et al. 1997). Crisis reactions of organisations, on the other hand, are singular manifestations characterised by a high degree of flexibility. As a rule, they are based on knowledge that has been newly generated under time-pressure, which organisations use in crisis situations to produce creative ad-hoc reactions through improvisation. These spontaneous adaptations are conceptually delineated from resource development processes with dynamic capabilities (Zollo and Winter 2002: 340, Winter 2003: 991).

Hence, if organisations respond to crises primarily with ad-hoc reactions, the question arises as to whether the dynamic capabilities become unusable in a crisis and therefore no longer have to be preserved (Helfat and Peteraf 2003). There is no agreement on this issue in the literature on dynamic capabilities, but this chapter will take the example of unions to show how the examined dynamic capabilities developed during the crisis.

**Methodology**

The empirical material for this study was acquired with the aid of semi-structured interviews conducted with trade union secretaries throughout Germany. The first phase of the survey took place in 2008 prior to the recession. The interview guidelines were adapted due to the changed situation resulting from the crisis in the second empirical research phase conducted in 2009.

In order to select relevant interview partners, firstly, statistics from the Federal Employment Agency were analysed in order to ascertain in which districts the temporary workers ratio was higher than the national average. Secondly, a survey of IG Metall detailing the temporary employees newly joining the union indicated where particularly successful routines had led to the recruitment of new members. Thirdly, respondents were asked about other potential interviewees who were known for their particularly active work in relation to temporary employees ('snowball principle').

Of the experts interviewed, 81 are responsible in IG Metall at a national (4), regional (7) and local (70) scale for the support of temporary employees. The contents of the interviews were recorded and transcribed.

One frequent criticism made about the concept of dynamic capabilities is that the question as to whether or not an organisation has dynamic capabilities cannot be answered until they are used and become manifest in practice as resource development processes (Helfat et al. 2007: 31). Numerous authors have therefore expressed concern that dynamic capabilities can only be identified through their effect (e.g. Eisenhardt/Martin 2000: 1114). I agree with this criticism, but argue that the actors in organisations are aware of their routines and organisational capabilities. They are able to cast them into words because they are characterised by verbal and written expressiveness, targeted use and reflexivity. Furthermore, changes to the resource basis occur consciously (Helfat et al. 2007: 1). Following the completion of qualitative interviews, the transcripts revealed statements by the interviewees about when which routines and organisational capabilities were consciously changed in the union and what learning processes were executed. These effects of dynamic capabilities, the conscious development of routines and organisational capabilities, could therefore be clearly identified in the data collected. The evaluation and interpretation of the interviews was performed by means of qualitative content analysis for this purpose (Mayring 2003).

## Adaptation Responses of IG Metall prior to and during the Crisis

The unions in the German Trade Union Confederation (DGB), of which IG Metall is a member, took a critical view of temporary work from the outset because it resulted in a decline in wage and social standards. From the mid-1990s, the DGB unions realised that they had to cease their demand for this booming and politically-supported type of employment to be prohibited. Since then recruitment efforts among temporary workers of IG Metall offices at various locations could be observed.

As a response to the increase of temporary employment, new resource constellations were developed from routines and capabilities with dynamic capabilities, which enabled the union to react to the rapid increase of temporary work within its environment. The newly developed organisational capabilities were those needed for recruiting temporary employees and for the representation of their interests. In addition, new routines were developed to exercise these organisational capabilities. The following empirical results will show which specific prerequisites triggered the development of resources prior to the crisis.

*Prerequisites for Resource Development prior to the Crisis*

The application of dynamic capabilities had begun because of a significant change in the environment of the union organisation, which fulfilled the aforementioned first prerequisite in this regard. From the mid-1990s, the federal government came to consider temporary work as a bridge to the employment market. In 2004 many legal restrictions were removed. As a consequence, between 2004 and 2008 temporary work underwent a sharp increase in Germany with the result that the number of temporary workers reached its current peak of 761,000 in 2008 (Federal Employment Office 2011).

However, these changes in the environment of IG Metall led the trade union secretaries in the organisational sub-entities of the union to resource development only if they resulted in an increase in temporary work within the area of responsibility. This was not uniformly the case among administrative offices or districts because the temporary worker rate exhibited distinct geographical differences. In rural areas temporary work is insignificant, whereas large urban agglomerations exhibit above-average numbers and rates of temporary employees. Because assignment firms are found in the automotive, mechanical engineering, electrical, steel, ship building and aeronautics industries, the locations of these industries likewise showed a high demand for temporary workers. For example the VW plant in Wolfsburg achieved the German-wide peak for a temporary employment ratio based on the total number of employees with 9.5 per cent (= 10,400 temporary workers) in June 2008 (Federal Employment Office 2011). Those union locations in whose area of responsibility temporary work rose to a high scale up to 2008 tended to become engaged more intensively on behalf of the interests of temporary employees.

Many union locations of this type also fulfil the second prerequisite for resource development. Promoters can be found there, i.e. trade union secretaries who react to these changes in their environment and accord quite a high priority to the representation of interests and recruitment of members for temporary workers.

The promoters are embedded in networks with actors in their area of responsibility, which fulfils the third prerequisite. This particularly relates to works councils in assignment firms. It is via these contacts that the promoters access knowledge and cooperate with the bodies representing the interests of temporary workers. One union secretary stressed: 'To gain access to the temps in a firm, it is indispensable to act jointly with works councils. In companies where works councils ignore the concerns of the temps or where no codetermination structures exist at all, you have no chance to get in contact with the temporary workers' [translation: D. Meyer]. For example, trade union secretaries together with works councils have invited temporary employees to meetings and offered to check the correctness of the salaries they receive from the temporary employment companies. In some cases, temporary workers subsequently joined the union and pursued legal means to demand the right to lawful payment of salaries. Through the cooperation between trade union secretaries and works councils a new, frequently empirically observed routine was developed. At the same time the new capabilities of representing the interests of temporary employees and recruiting them was established.

The fourth prerequisite for the application of dynamic capabilities is fulfilled if favourable opportunities arise to improve the working conditions and wages of temporary workers. If, for example, the management of an assignment firm wishes to conclude an operating agreement with the works council, the unions support or advise these works councils during the negotiations. In a total of 400 operating agreements that were concluded in 2008 alone, the works councils successfully pushed through salary improvements for the temporary workers. Through the joint efforts of the works councils and union secretaries another new routine has been established within the IG Metall.

Because changes within the union organisation also stimulated some promoters to undertake activities, the fifth condition for the use of dynamic capabilities was also fulfilled. Since the 1980s, the membership numbers of the DGB unions have been in decline (DGB 2011). In order to assert political influence, to regain acceptance within society and to halt the decline of financial resources, the recruitment of new members is one of the most pressing aims of the unions. This pressure has caused the promoters to develop new resource configurations with dynamic capabilities. These interconnected bottom-up and top-down resource development processes that subsequently took place, is described below, taking IG Metall as an example.

*Resource Development Bottom-Up*

In the mid-1990s, a hostile attitude to temporary work prevailed among most secretaries in the union organisation, being denunciated as slave work. However, at some union locations where the aforementioned prerequisites were fulfilled, a bottom-up process was put in motion, initiated by the promoters on the local union scale. These promoters took the initiative to develop resources in learning processes with dynamic capabilities.

Through learning-by-searching, the promoters identified which companies within their area of responsibility were using temporary workers. One of the interviewed secretaries commissioned a research institute to perform a survey to quantitatively ascertain the local extent of temporary employment. Using another learning process, learning-by-interacting, the promoters in the administrative office at Siegen exploited their well-established, close contacts with works councils in local temporary employment agencies in order to access knowledge about temporary workers and their wage and employment situations. Through experimental learning the promoters then tested new methods in practice, which, if successful, they concretised into routines. However, such initiatives did not always actually lead to the desired success. A union secretary, who organised regular meetings with works councils and temporary employees in the administrative office for the purpose of information exchange, ceased these efforts due to a lack of participation.

Besides this when new opportunities to establish innovative routines arose, the promoters seized the opportunity and successfully generated a new routine. In Regensburg, temporary employees turned to the promoter in order to establish a works council in their temporary employment agency. The secretary established local co-determination structures within the temporary employment firm. The promoter even succeeded in repeating this in another temporary employment agency so that we can talk of a locally established routine in this instance. During the foundation of the second works council, the promoter was able to exploit previously acquired insights to implement the routine and supplemented this knowledge with new experiences. Therefore, the second instance was no longer a case of experimental learning but one of learning-by-doing.

The local learning processes of learning-by-searching, learning-by-interacting, experimental learning and learning-by-doing consequently resulted in the establishment of new routines. The organisational capability to represent interests and recruit members among temporary workers gradually matured on the local scale.

At union meetings, promoters highlighted the issue of temporary work, the resulting problems of the dramatic increase in the significance of temporary work were discussed and collective learning processes took place when the promoters reported about their efforts towards recruiting members from the temporary agency employment sector. Most secretaries on the local, regional and national scale were interested in how union secretaries at some locations had already successfully implemented new routines and capabilities for handling the potential new membership group. As the discussion and knowledge-transfer about the new direction of union policy vis-à-vis temporary work intensified in large parts of the organisation and the new knowledge from the local scale found wider acceptance within the organisation, more and more administrative offices participated in the use of dynamic capabilities and the subsequently developed their own routines. One of the interviewees was given advice by his colleagues operating at the local level as well, as he describes: 'I asked them: 'How did you get the works councils involved?' They had previously made up their minds about this question and we took advantage of their experiences' [translation: D. Meyer]. Thus, the new routines were disseminated at the local level of the organisation.

*Resource Development Top-Down*

In 1999, an EU Directive was enacted and determined that temporary workers should receive the same salary as permanent employees. Deviations were permitted however if collective agreements were in place. When, in 2004, German legislators enacted the new arrangement of temporary agency as part of the temporary employment law reforms, the equal pay principle was adopted into German law. In order to circumvent this, employers within the temporary agency employment sector pushed the conclusion of collective agreements. The DGB unions for the first time concluded collective agreements with temporary employment associations. The negotiation of collective agreements constituted a traditional routine of the union board, which were then further developed with the dynamic capability.

In 2007, the board of IG Metall in the district e.g. of North Rhine-Westphalia created new positions for secretaries who solely attended to temporary work. These positions were staffed with promoters who already possessed extensive knowledge of temporary work. They provide support, for example, to the works councils in assignment firms in their regional area of responsibility during the negotiations of operating agreements, in which it was stipulated that only temporary workers from reputable temporary employment agencies may be used.

While some local promoters, due to their efforts, were able to record successes in relation to the recruitment of members, the board intensively discussed the realignment of the complete organisation towards temporary work. One significant event that decisively influenced the future course of IG Metall was the union congress held in 2007. In order to establish a nationwide uniform approach towards representing the interests of temporary workers and recruiting them, the board promoted the resolution of a temporary employment campaign and received widespread support at the local and regional scales. The main objectives of the campaign included the implementation of the principle of equal pay as well as the achievement of a high rate of union organisation within the temporary agency employment sector. The routines implemented included training works councils to achieve professional standards in supporting and recruiting temporary employees within the assignment firm. Those union secretaries on various local union posts who had not yet become active were offered support. According to a union secretary who co-managed the campaign, the union board requested all local union administrations to join its campaign: 'The goal of our campaign is to address temporary workers in each and every single local and regional administration in Germany. From north to south, from east to west. We pool all activities and thus gain momentum' [translation: D. Meyer].

Thus, the object of the campaign was to disseminate the routines and organisational capabilities top-down within the organisation and to implement them within all sub-entities. By these means the new knowledge base and resource configurations, generated and developed on local and regional scales, have an impact across the organisation. The application of the dynamic capabilities, moreover, should be performed throughout the organisation and new learning processes encouraged.

As a result, there was an increase in the actual nationwide commitment of IG Metall secretaries to the interests of temporary employees. The fact that the dynamic capabilities were applied at numerous union locations to successfully recruit members up to the crisis is demonstrated by the 10,000 temporary workers who joined IG Metall in 2008.

As the statistics on membership recruitment show, some locations remained inactive. Reasons mentioned include that temporary work had no significance for the local employment market.

*Dynamic Capabilities during the Crisis*

With the onset of the economic crisis in late 2008 there was a change in the prerequisites for the development of resources:

First, the crisis manifested itself in the external environment of the union organisation by a geographically diverse decline in the number of temporary workers. In July 2008, the number of temporary workers in Germany achieved its highest monthly figure to date, with some 823,000 employees. By April 2009, mass lay-offs led to a decline in the numbers of temporary workers to 580,000, its lowest point during the crisis (Federal Employment Office 2011). The most drastic cuts in staff were implemented in particular by temporary employment companies in regions in which previously industrial companies had demanded temporary workers. As a consequence, for example, the automotive and engineering hub of Stuttgart and the ship building and aerospace centres in northern Germany were strongly affected by the crisis. Temporary worker numbers remained stable in places where temporary employment firms found alternative deployment options in service sector branches less affected by the crisis.

Second, most union secretaries responded to the decline in temporary employment by suspending the application of the dynamic capabilities and did not use the new resources any longer. Only the promoters of very few administrative offices perceived an especially great need for action precisely because of the conditions imposed by the crisis. Their new objective was to prevent the lay-off of temporary workers.

Third, the majority of the works councils in those assignment firms that had previously acted on behalf of temporary workers ceased their activities. Because during the crisis more importance was attached to representing the interests of full-time employees within companies, they were no longer available to promoters in unions as partners for action.

Fourth, new opportunities also arose in the environment during the crisis. These were only exploited by those promoters who had remained active. For example, it was frequently the case during the crisis that the management of a company that had utilised temporary work prior to the crisis wanted to conclude an alternative collective agreement with the local union concerning the working hours and salary of the permanent staff. Therefore, an interviewed secretary explained: 'This placed us in a favourable negotiating position in turn to push for agreements on regulations concerning temporary work. Looking ahead to the end of the crisis, we utilised the crisis during which no temporary workers were being hired in firms, to establish better conditions for future temporary worker deployments' [translation: D. Meyer]. The promoters could, for example, demand that every six months a certain number of deployed temporary workers would be offered a fixed employment contract.

Fifth, the pressure to act grew within the union. Efforts had to be made to ensure that union members among temporary workers who had become unemployed did not leave IG Metall. In addition, the long-term union members who had previously had permanent employment with industrial enterprises and who had become unemployed during the crisis now urgently required the support of the local trade union secretaries.

Under the new conditions the question arose whether the dynamic capabilities could continue to be used for the development of routines and capabilities by way of an adaptive response. The empirical results indicated that while the dynamic capabilities were indeed less frequently and actively used, they nevertheless continued to exist.

On the local scale they continued to manifest themselves in the form of newly developed routines and capabilities. In June 2008 in Freiburg, for example, at the initiative of a promoter a working group was formed by temporary workers and works councils to tackle issues relating to temporary agency employment. The commitment of these actors did not falter following the onset of the crisis.

According to the information provided by the interviewees, on the regional scale the majority of districts did indeed respond to the declines caused by the crisis by reducing their activities to a minimum scale because the personnel capacity was required in other fields of activities. In few districts even during the crisis, the temporary employment representatives continued to develop new routines and to support the administrative offices in their region. The secretary in the North Rhine-Westphalia district, for example, informed the local secretaries through seminars about fraudulent dismissal practices used during the crisis by temporary employment firms that failed to observe legal dismissal regulations or stopped paying salaries.

On the national scale the board drastically reduced its top-down implementation of the campaign but still remained active. In the summer of 2009, for example, it organised a convention for local administrative offices and works councils from throughout Germany to exchange knowledge concerning temporary employment. In addition, the union board renegotiated the collective agreement provisions with the associations of the temporary agencies.

Consequently, the dynamic capabilities were not unlearned during the crisis but instead were less intensively applied for a temporary period. This is supported by the fact that their importance rose again within the organisation following the end of the crisis.

However, the reactions of the union cannot be explained with dynamic capabilities alone. It was only in combination with ad-hoc reactions that they led to adaptations. A few of the union locations developed ad-hoc reactions enabling them to rapidly adapt to the crisis.

For example, the IG Metall board campaigned with the federal government to also apply the short-term working provisions to the temporary agency employment sector so that temporary agency employment companies could also apply to receive short-term employee benefits. Short-term employee benefits are financial assistance provided by the Federal Board of Employment. Companies may avail themselves if they are suffering from a temporary downturn. In the most extreme cases, employees will not perform any work at all but the Federal Board of Employment will nevertheless continue to pay those affected up to 67 per cent of their lost net income during this period. To lobby for the new law concerning short-term work for temporary workers is not to be regarded as a new, repeatedly applied routine but instead as a spontaneous ad-hoc reaction implemented immediately after the start of the crisis.

Improvised adaptation responses of this kind were also observable on the local scale. In December 2008, the Cologne administrative office was successful in preventing the dismissal of 260 employees of the temporary employment company Adecco who were deployed in the Ford assembly plant. Adecco wanted to lay them off because Ford had no further use for them. Local trade union secretaries brought the relevant parties together for discussions to find another solution by which the jobs of the temporary workers were preserved. Short-term work was requested for the temporary workers and they were able to participate in state-financed training programmes.

There are no precise figures to indicate the extent to which it was possible to prevent the crisis from decimating membership recruitments made among temporary workers prior to the crisis. According to the information provided by the interviewees, however, there were no immediate losses of newly recruited temporary workers. Furthermore, contrary to expectations, the membership losses of IG Metall overall were kept to a low level during the crises years of 2008 and 2009. This indicates that – at least in some locations – the dynamic capabilities and ad-hoc reactions proved successful.

**Conclusion**

From the mid-1990s, some IG Metall union secretaries identified temporary workers as a new target group for the recruitment of members and started to represent their interests. It was shown how these secretaries first had to learn the requisite organisational capabilities and routines to be able to do this. It were their dynamic capabilities that, prior to the onset of the economic crisis in late 2008, enabled the IG Metall union organisation to respond to the drastic increases in temporary work by taking action for the improvement of wages and working conditions of temporary workers.

Since the union organisation is divided into various sub-entities, identical resource configurations were not created throughout the entire organisation. Instead, differing scales of activity led to the development of various routines and organisational capabilities at the respective union locations (multi-location) and across the three scales of the union organisation (multi-scalar).

The prerequisites for resource development with dynamic capabilities were highlighted in this chapter: First, the increase of temporary work within the environment; second, the activities of promoters; third, the cooperative efforts undertaken with works councils in assignment firms; fourth, the creation of new opportunities arising from the action of the management of assignment firms. If the management wanted to negotiate new operating agreements with works councils or collective agreements with the union, it was possible in turn to demand concessions in favour of temporary workers within the company. The influence of management union activities is stressed as well by Heery and Simms (2010).

Finally, the fifth prerequisite mentioned is the internal pressure to act, which resulted from the union's membership losses.

Next, the concrete learning processes that will occur if the prerequisites are satisfied were described: Learning-by-searching and by -interacting, experimental learning and learning-by-doing. The new routines and organisational capabilities are disseminated bottom-up within the organisation through communicative exchanges, before being ultimately implemented through a temporary work campaign by the board in a systematic top-down fashion extending almost entirely throughout the organisation.

With regard to the learning processes, a different significance is attached to each of the three scales. Due to the multi-scalar and multi-locational structure of the union, the bottom scale is ascribed high importance, but possesses little potential for disseminating new concepts throughout the organisation. The regional scale of the districts is vital because it transfers the impulse from the administrative offices to the executive scale. To achieve change throughout the union, the union board on the national scale fundamentally influences the general course.

When the conditions changed following the onset of the crisis, new adaption reactions had to be established. The decline in the number of temporary workers as well as the abatement in the commitment of many union secretaries and works councils within assignment firms meant that the dynamic capabilities were applied less intensively. However, in those places where promoters remained active it was possible to seize new opportunities emerging within the assignment companies, with the result that the dreaded loss of union members was averted. Under the changed conditions, the dynamic capabilities consequently were utilised for the multi-scalar and multi-location development of routines and capabilities as an adaptive response – even if the degree of utilisation was less intensive and more sporadic than previously. The dynamic capabilities were supplemented across the three scales by ad-hoc reactions which equip organisations with particularly rapid and flexible adaptive responses that they require in times of industrial crisis.

## References

Argyris, C. and Schön, D.A. 1978. *Organizational Learning*. Reading: Addison Wesley.

Barney, J.B. 1991. Firm Resources and Sustained Competitive Advantage. *Journal of Management*, 17(1), 99–120.

Bundesagentur für Arbeit [Federal Employment Office] 2011. *Zeitarbeit in Deutschland. Aktuelle Entwicklungen [Temporary Work in Germany. Current Developments]*. Nuremberg: Bundesagentur für Arbeit.

Coe, N.M., Johns, J. and Ward, K. 2009. The Temporary Staffing Industry and Labour Market Restructuring in Australia. *Journal of Economic Geography*, 9(1), 55–84.

Cumbers, A., MacKinnon, D. and Shaw, J. 2010. Labour, Organisational Rescaling and the Politics of Production. *Work, Employment and Society*, 24(1), 127–44.

Deutscher Gewerkschaftsbund (DGB) [German Trade Union Confederation] (2011): *Die Mitglieder der DGB-Gewerkschaften [The Members of the DGB Unions]*. [Online] Available at: http://www.dgb.de/uber-uns/dgb-heute/mitgliederzahlen [accessed: 11 October 2011]

Eisenhardt, K., M., Martin and Jeffrey A. 2000. Dynamic Capabilities: What are They? *Strategic Management Journal*, 21(10–11), 1105–21.

Gertler, M.S. 2003. Tacit Knowledge and the Economic Geography of Context, or the Undefinable Tacitness of Being (There). *Journal of Economic Geography*, 3(1), 75–99.

Frege, C.M. and Kelly, J. 2003. Union Revitalization Strategies in Comparative Perspective. *European Journal of Industrial Relations*, 9(7), 7–22.

Heery, E. and Simms, M. 2010. Employer Responses to Union Organising: Patterns and Effects. *Human Resource Management Journal*, 20(1), 3–22.

Helfat, C.E. and Peteraf, M.A. 2003. The Dynamic Resource-Based View: Capability Lifecycles. *Strategic Management Journal*, 24(10), 997–1010.

Helfat, C.E., Finkelstein, S., Mitchell, W., Peteraf, M.A., Singh, H., Teece, D.J. and Winter, S.G. 2007. *Dynamic Capabilities: Understanding Strategic Change in Organizations*. Malden, Oxford, Victoria: Blackwell Publishing.

Herod, A. 2010. Labour Geography: Where Have We Been? Where Should We Go?, in *Missing Links in Labour Geography* edited by A.C. Bergene, et al. Surrey, Burlingston: Ashgate, 15–28.

Lévesque, C. and Murray, G. 2010. Local Union Strategies in Cross-Border Alliances. *Labor Studies Journal*, 35(2), 222–45.

Levitt, B. and March, J.G. 1988. Organizational Learning. *Annual Review of Sociology*, 14(1), 319–40.

Mayring, P. 2003. *Qualitative Inhaltsanalyse [Qualitative Content Analysis]*. Weinheim: Beltz.

Nelson, R.R. and Winter, S.G. 1982. *An Evolutionary Theory of Economic Change*. Cambridge, London: Belknap Press.

Peck, J. A. and Theodore, N. 2007. Flexible Recession: the Temporary Staffing Industry and Mediated Work in the United States. *Cambridge Journal of Economics*, 31(2), 171–92.

Tattersall, A. 2006. *A Framework For Analysing When Unions Are Likely to Practice Collaboration with the Community*. University of Sydney: Working Paper, No. 23.

Teece, D.J., Pisano, G. and Shuen, A. 1997. Dynamic Capabilities and Strategic Management. *Strategic Management Journal*, 18(7), 509–33.

Teece, D.J. 2009. *Dynamic Capabilities and Strategic Management*. New York: Oxford University Press.

Wiesenthal, H. 1995. Konventionelles und unkonventionelles Organisationslernen [Conventional and Unconventional Organisational Learning]. *Zeitschrift für Soziologie*, 24(2), 137–55.

Wills, J. 2005. The Geography of Union Organising. *Antipode*, 37(1), 139–159.
Zollo, M. and Winter, S.G. 2002. Deliberate Learning and the Evolution of Dynamic Capabilities. *Organization Science*, 13(3), 339–51.

Chapter 7

# Industry Transition and Knowledge-Based Sectors: Changing Economic Governance and Institutional Arrangements in the Scottish Life Sciences

Kean Birch and Andrew Cumbers

## Introduction

Developed economies face an uncertain future in light of the ongoing financial crisis. The reliance on personal consumption, rising asset prices and financial services as drivers of economic development has proved to be misplaced, to say the least. The choices facing developed countries are, however, restricted by previous decisions and the resulting circumstances these have engendered. Consequently of the kinds of industrial transition open to policy-makers may be limited, as they seek new ways to position developed economies within the wider world. Changing global imperatives have driven many countries towards particular forms of industrial transition, pursuing particular policies in order to stimulate, encourage and support certain types of industries. These have been based on the perception that developed economies are increasingly 'knowledge-based', 'knowledge-driven' or some similar epithet, espousing the truism that all human activity involves the processing and application of knowledge, learning and innovation.

The focus on knowledge, learning and innovation represents a particular facet of industrial transition in which global 'competitiveness' is promoted through the expansion of knowledge-based sectors like the life sciences. Mainstream debates have centred on several issues with regards to this transition to an emerging *knowledge-based economy* (KBE). However, there is increasing fuzziness between the positioning of developed and developing countries along and within global commodity chains and production networks. It is by no means inevitable that developed economies will be able to capture the high-end activities in knowledge-based sectors, especially as developing economies expand their investment in these areas.

The emerging global shift towards knowledge-based sectors has an important local dynamic as well. In particular, many less-favoured regions (LFRs) in developed economies are threatened by the shift to knowledge-based sectors on top of the loss of manufacturing to developing countries. These LFRs often lack the appropriate human and technological assets of more dynamic regions to make the transition upwards in production and innovation processes to higher value-added activities. A key element of policy has therefore been to seek to upgrade the regional knowledge assets of such regions to cope with the changing competitive conditions of the global economy. However, it has been argued that a focus on the regional scale itself is unlikely to be enough, meaning that policy-makers and others need to think about the repositioning of regions within broader knowledge networks (MacKinnon et al. 2002). It is our argument here that successful industrial transition in knowledge-based sectors can only be secured through, on the one hand, a reconfiguration of economic governance along *global* commodity chains and, on the other hand, the grounding of these commodity chains within particular institutional arrangements at the local and regional scales.

In this chapter, we explore these issues and debates through a case study of the Scottish life sciences, which we argue represents a relatively successful example of industrial upgrading and transition in an LFR. This success story has been achieved against a backdrop of long-term and indeed continuing decline in traditional manufacturing sectors, in conjunction with a long history of failed regional policy initiatives aimed at securing adjustment and transition along new development pathways. However, the ability to sustain this success story in the future hinges upon the ways that Scottish life science firms are being plugged into the broader global geography of the life sciences.

## Industrial Upgrading in Knowledge-Based Sectors

*Global Commodity Chains and Knowledge-Based Sectors*

One way of exploring industrial transition is through a global commodity chains (GCC) approach. There is an enormous amount of research on GCCs and there is not the space to fully review this literature here (see instead Bair 2005). The concept itself can be traced back to the Hopkins and Wallerstein's (1986) world-systems theory, which was explicitly concerned with the chain of labour and production processes that underlie commodity exchange. The seminal work of Gary Gereffi (1994) in this field highlighted several key benefits of the GCC approach. However, there are a number of issues with applying the GCC approach to knowledge-based sectors, which we seek to address in this chapter and elsewhere (see Birch 2008, 2011, Birch and Cumbers 2010).

The main issue is that the GCC approach has relatively little to say about economies where knowledge and other intangible assets represent an increasingly dominant – if not already dominant – proportion of economic activity. Although the GCC approach has always been concerned with knowledge, it has focused on technological upgrading in developing countries rather than in the less-favoured or peripheral regions in developed economies. In order to examine knowledge-based sectors, therefore, it is important to go beyond the existing GCC approach to look at production, labour *and*, most importantly, innovation processes behind *knowledge-based commodity chains*.

Current research trends in the GCC field do not provide much of an avenue for this task, however. For example, Gereffi has recently theorised a global value chain (GVC) approach with others (see Gereffi et al. 2005), but this recent concept only focuses on the relationship between two actors within a chain (Bair 2005). It therefore (a) provides a limited purchase on the global labour, production and innovation processes along the commodity chain, and (b) ignores the specific, local institutional geographies underpinning innovation at different points of the commodity chain. It is crucial to consider this local-global dynamic in order to understand the geographies of industrial transition; that is, the positioning and re-positioning of firms along the commodity chain.

In attempting to advance our understanding of the economic governance and institutional embedding of processes of industrial upgrading in knowledge-based sectors, it is important to adopt an approach that reflects the flatter and less hierarchical networks of innovation and production in which such knowledge-based sectors operate (see Birch 2008, 2011, Birch and Cumbers 2010). Unlike earlier GCC models, any analytical approach must reflect the essentially non-linear realities of knowledge production such that traditional economic development concerns with upgrading are replaced by attention to the changing dynamics of industrial transition and their embeddedness within global *and* local networks. In this sense we account for two important analytical points that are currently under-theorised in the GCC literature and helps position debates on industrial transitions within the wider literature. First, the economic governance of different priorities and agendas is not necessarily dependent upon localised relations or assets alone. Second, the economic restructuring of knowledge-based sectors remains embedded in an array of institutional arrangements, some local and some not.

## Governance and Trust in Industrial Upgrading

The nature of trust and its spatial implications are of interest when it comes to industrial transition. In existing research on knowledge, space and economic relations, trust has been viewed as an important asset underpinning economic performance, especially in knowledge-based activities that require considerable interaction and collaboration between actors and where, departing from mainstream economic analysis, knowledge is always assumed to be partial (Storper 1997).

Discussions of trust have tended to focus upon it as a localised phenomenon, bound up in endogenous business and development networks, which when it operates successfully provides considerable 'untraded interdependencies' for regions and small firms in local business clusters (Storper 1995). Researchers have found it useful to distinguish between 'competence trust', where collaborators can be confident (usually through reputation and past performance) that their business partners have the ability to carry out certain activities to a required standard, and 'goodwill trust' which refers to the a deeper level of trust, often tied up in personal or social relations that go beyond a narrowly defined business dividend and are present in longer term networks and associations (MacKinnon et al. 2004).

As is now widely recognised in the literature, trust is not a localised phenomenon but is critical to the functioning of business networks, especially in knowledge-based sectors (MacKinnon et al. 2004). At the regional scale, successful industrial transition entails the intersection of diverse communities of practice, which requires both kinds of trust developed through more open networks of often global knowledge exchange and intellectual endeavour alongside the need for reliability in delivering a product to the marketplace. Tensions necessarily rise in these trust relations between the need for openness in the exchange of ideas and the economic imperatives to capture value through the protection and definition of intellectual property rights. Furthermore, the complexity of knowledge-based commodity chains, as in the life sciences, entails a further dimension of trust that is related to the need to meet global regulatory standards and protocols (e.g. International Standards Organisation). This involves the *objectification* of relationships, where it is the adherence to the global standards themselves that engenders trust rather than the relationship itself. We conceptualise this as *objective trust* standing in contrast to the inter-subjective forms of trust outlined above. Such objective trust is embodied in objects and practices that adhere to particular operational standards (e.g. clinical, manufacturing), which ensures that the trustworthiness (i.e. 'quality') of those objects and practices is bound up with the process of standard-setting rather than previous performance or personal connections.

The potential implications of trust for successful industrial transition are profound for they suggest that knowledge-based sectors may be subject to very different imperatives to those that structure spatial divisions of labour in other sectors (Massey 1995). Successful product development requires collaboration and the bringing together of diverse and complex knowledges across organisations and space in a dynamic and uncertain market place that defies attempts to dominate and centralise power but instead requires the exercise of very different spatial modalities of governance to facilitate the kinds of collaboration and trust identified here. In particular, regional upgrading strategies require the development of more associational forms of governance and trust, embedded in broader institutional geographies that we turn to next.

## The Institutional Grounding of Industrial Transition

With the 'institutional turn' in economic geography, there has been a significant growth in the literature relevant for understanding the social and institutional context of industrial transition (e.g. Amin and Thrift 1992, Morgan 1997, Cooke and Morgan 1998). Largely built around the sociological idea of *embeddedness*, this literature emphasises that economic actions and landscapes are thoroughly entangled in social relations, an idea which has, in turn, underpinned the 'new regionalism' agenda that is implicated in the decentralisation of political-economic decision-making (Pike et al. 2006). This literature and associated political agendas highlights the role that a location's social and political conditions (or assets) play in encouraging and supporting – or obstructing – new industrial sectors like the life sciences.

Such local institutional conditions include language, social norms and conventions, culture and shared expectations, all of which are seen as generating trust thereby contributing to ongoing cooperation, collective learning, networking and the flow of knowledge between organisations and agents in particular localities (Gertler and Levitte 2005). In these conceptual approaches, particular places are characterised by a self-reinforcing institutional environment (e.g. social conventions, culture etc.) accompanied by more formal institutional arrangements (e.g. labour market, universities etc.) that enable them to better adapt to economic change (Pike et al. 2006). With regards to knowledge-based sectors, such as the life sciences, these institutional arrangements have an important (and dynamic) impact on the organisation and governance of industrial transition through the inherited social conditions that influence how particular places can respond to the imperatives of globalization (Cumbers et al. 2003).

## Exploring Industrial Transition in the Scottish Life Sciences

The life sciences have been one of the few instances of economic success and renewal in Scotland against a broader background of economic decline in the country's traditional manufacturing industries. Scotland represents an interesting case because it is a 'region' of the UK that has suffered from severe de-industrialisation and uneven development over the last thirty years, losing around 40 per cent of its manufacturing employment since the mid 1990s (Birch et al. 2010). Until recently, large parts of central and lowland Scotland were deemed worthy of considerable European regional assistance because of poor economic performance and social deprivation. It is therefore a 'region' that can be considered 'less-favoured', even though some places such as Edinburgh and Aberdeen have grown strongly over the last three decades. The life sciences represent a sector that offers potential as a knowledge-based transition to new forms of industry and employment.

Altogether the life sciences industry employs around 30,000 people across 600 private and public sector organisations (including the Universities of Aberdeen, Dundee, Edinburgh and Glasgow) making Scotland the most significant 'cluster' of life sciences activity in the UK outside London and the south east of England (Birch 2009). Our empirical work is based upon a study of the Scottish life sciences carried out in 2008 using a multi-method research design incorporating a number of different stages. The first stage involved mapping the 600 organisations included in the life sciences using Scottish Enterprise's 'Sourcebook' database (see Figure 7.1). As Figure 7.1 illustrates, the life sciences are concentrated in the central belt of Scotland between the major cities of Glasgow, Edinburgh and Dundee with smaller sub-clusters around Aberdeen and Inverness. Afterwards we surveyed all the 'core' life science firms ('core' defined as those firms using biological techniques and applications to develop products or intellectual property). Using

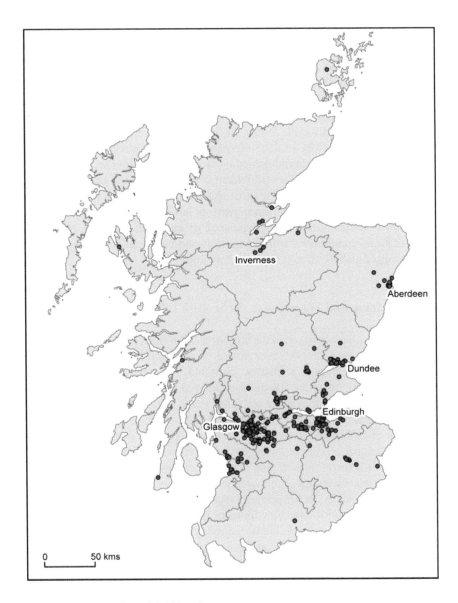

**Figure 7.1     The Scottish life sciences**

this definition, we identified 190 firms from which we carried out a telephone questionnaire with a 39 per cent response rate (n = 74).

The emergence of the life sciences owes much to a long tradition of bio-medical research in Scottish universities (extending back to the eighteenth century Enlightenment), allied to public sector support for medical research dating back to the 1940s.  However, most of the firms we surveyed have been established in the last two decades: 39 per cent in the 2000s and another 33 per cent in the 1990s. This provides encouraging signs that less-favoured regions like Scotland going through industrial transition are not destined to lose out on the economic benefits of new knowledge-based sectors.

This example of successful industrial transition is based on sustained collaboration between key actors in government, universities and the business sector; all of whom have contributed to the emergence of the life sciences as a new, dynamic and global sector in Scotland. While these 'regional' institutional arrangements have been crucial, the growth of the life sciences has been facilitated by wider spatial networks from the outset. Of particular importance are the global knowledge communities within which Scottish scientists and academics operate, and a considerable Scottish scientific diaspora which is reproduced by flows of labour and knowledge of varying durations to and from 'the region'. Added to this, the geography of the life sciences market, in terms of the considerable economies and technologies of scale and the subsequent dominance of large transnational corporations, means that Scottish firms must be 'global' from the outset in selling their products and ideas.

The second stage of the project comprised interview with 32 key actors in the Scottish life sciences. First, we interviewed 19 informants from 18 life science firms. Second, we interviewed 13 'institutional' actors drawn from various Scottish organisations connected to the life sciences (e.g. Scottish Government, trade associations, business angels, industry advisory bodies etc.). The focus in the remainder of this chapter is on our findings from these interviews; we don't cite from these interviews directly for space reasons. However, see Birch and Cumbers (2010) and Birch (2011) for fuller discussions of each aspect of this project. First, we explore the governance and coordination of activity along the commodity chain and highlight, how power plays out between different actors and what impact this has on regional economic restructuring. And, second, we examine the institutional arrangements in which Scottish life science firms operate.

## The Economic Governance of Industrial Transition

*Governance, Coordination and Trust*

The complexity of the life sciences is evident in the intricacy and variety of relationships along their knowledge-based commodity chains. As argued in the theoretical section, industrial upgrading and transition entail new forms of economic governance driven by global alliance-making and collaborative competencies in which the coordination of specialised and interdisciplinary capabilities and incentives across different organisations is paramount (Birch 2008).

First, the complexity of new technoscientific discoveries and innovations means that no single organisation – whether multi-sited, multi-national or not – is capable of integrating all the necessary competencies in their internal organisational structure. Innovation is a necessarily dispersed process that cannot be concentrated or captured in one place, or by a single set of actors, meaning that there is always a need for active and global searching for complementary knowledge.

Second, there is an evident shift in the form of knowledge production entailing interdisciplinary, transdisciplinary and multidisciplinary working since life sciences firms engage in diverse areas of research. Knowledge production, for life science firms, entails global interaction and searching across different disciplines (e.g. anti-inflammatory, glycobiology). In the Scottish life sciences, for example, firms plug into wider, global knowledge networks which provide them with access to a wider array of expertise and know-how than might not be available in Scotland alone.

As the recent emphasis in the literature on global pipelines (Bathelt et al. 2004) has emphasised, such broader spatial connections are critical for economic restructuring and, successful regional development. But the main point to emphasise is that it is imperative to retain open, dynamic and global linkages in knowledge-based sectors to retain access to knowledge networks. Ultimately the expertise and knowledge necessary for life science innovation cannot be located or remain concentrated in any one organisation – or even one region – because so many different disciplines, capabilities and assets contribute to the innovation process at the same time that the risk for any one organisation of attempting integration is too great. The dynamic nature of knowledge within the sector thus encourages more dispersed geographies of innovation. The result is that not only will the types and forms of knowledge be different across different organisations and regions, but the processes of learning, upgrading and commercialisation will vary at the same time depending on a variety of factors such as – but not limited to – the size, type and reach of each organisation. In this sense, industrial transition in knowledge-based sectors like the life sciences is driven by the positioning and repositioning of firms along *global* commodity chain. Economic governance of this (re-) positioning is necessarily tied to specific geographies of coordination and trust in that firms cannot operate alone or *in situ* nor without reference to broader institutions and market influences.

*Governance of Knowledge Relations in the Life Science Commodity Chain*

As we have emphasised, Scottish life science firms engage in a variety of relationships along global commodity chains. These linkages, however, cannot be considered as homogenous because they involve a variety of organisations, different institutional settings, different types of relationship, different geographical dimensions and so on. All these linkages entail different geographies of governance in that they involve complex ways of coordinating relationships across different locations.

First, the governance of knowledge relations is underwritten by the *objectification* of trust-based relationships, which contrasts somewhat with the conceptualisations of 'competence' and 'goodwill' trust discussed in the theoretical section above (see

Cumbers et al. 2003, MacKinnon et al. 2004). Despite the existence of these forms of trust and the ubiquitous reliance on personal recommendations in the Scottish life sciences, however, there is a strong emphasis on the objectification of firm-level capabilities and activities, which helps to produce a form of *objective trust*. This type of trust contrasts with inter-subjective forms of trust, such as competence and goodwill trust. For example, ISO standards come to represent the basis for the 'good relationships' that firms have with their suppliers. Notably, it is not the type or form of the relationships themselves that matters, nor is it the closeness of the relationships, nor the previous performance of the firm in question that engender trust. Rather it is the adherence to a particular set of standards embodied in objects and practices that engender trust.

Second, trust is an iterative effect of each firm's need to be compliant with global regulations (e.g. to ISO standards). Compliance then drives their need to ensure the regulatory compliance of their suppliers, manufacturers etc., wherever they may be. Since product development is such a long process, it is perhaps inevitable that firms build these 'objective' standards into their internal processes from an early stage, which has a recurrent, trust-engendering effect as firms seek 'quality assurance' along global commodity chains. So, whilst knowledge-based commodity chains are characterised by dispersed relations of innovation on the one hand, they are also characterised by adherence to global regulatory standards. This, in turn, engenders trust in their operations, which covers not only tangible product development but also knowledge production. The incorporation of these standards into a firm's practices means that other firms can trust the work that the firm is doing.

The issue of global regulatory compliance provides a good example of the complexity involved in the governance of knowledge relations; it illustrates how the coordination of different capabilities and incentives between organisations is structurally produced and reproduced along the commodity chain. The social processes that constitute this coordination are built around the idea of 'objective trust', which both objectifies certain forms of relationship as trust-worthy – i.e. they are treated as a property of an object rather than social relations – and iteratively informs the practices of life science firms themselves as the 'production' of trust in their own activities necessitates the incorporation of standardised and approved procedures, often set by international organisations, that are accepted in different countries around the world. Consequently, such trust is not constituted by local or even national interaction, linkages and relationships; rather, as a mechanism bound up with the universalising tendencies in global regulatory regimes, it facilitates the convergence of practices across different geographies.

*Governance of Risk Outsourcing in the Life Sciences*

Whereas the governance of knowledge relations involves the (iterative) incorporation of global standards into firm-level practices, the governance of risk outsourcing involves both the embedding of differentiated knowledge in firm-level expertise and capabilities, and the protection of this knowledge in-house and when collaborating with external partners. Without the protection of in-house expertise there would be no incentive for other organisations to collaborate with a firm since the knowledge would be freely available. Whilst this may sound like it is detrimental to large multinational corporations, the production and protection of knowledge by small life science firms actually has a beneficial effect in that it helps to shift risk back down the commodity chain away from multinational corporations. This is because product development in knowledge-based sectors like the life sciences is extremely uncertain and therefore risky for large firms to invest in. So outsourcing risk involves social processes of coordination of organisational agendas and strategies that, again, entail particular geographies of economic governance along the commodity chain.

First, the governance of risk outsourcing is based, to a greater extent than knowledge exchange, on inter-subjective forms of trust such as 'competence trust' and 'goodwill trust' (see Cumbers et al. 2003, MacKinnon et al. 2004). For example, Scottish life science firms commonly relied upon the personal networks of their chairpersons or boards to recruit personnel, to access finance, to make contacts and so on. Furthermore, due to the particular out-licensing business model pursued by most life science firms, the relationship between small firms and larger ones was characterised by the coordination of specific incentives which serve to engender trust through the aligning of their different interests. One such mechanism is the use of royalties, rather than outright sale, as part of licenses and partnerships. Thus, in the light of the uncertain commercial viability of most firms' knowledge assets, the licensing model depends on the *performance* of confidence in one's own knowledge and expertise by accepting specific forms of remuneration; e.g. royalties instead of direct payments.

Second, such governance is intrinsically tied up with the protection of intellectual and intangible assets; that is, a firm's knowledge and capabilities. Without such protection there would be few opportunities for small firms to collaborate with larger firms and move up the commodity chain. What is striking about this process is the extent to which intellectual property (IP) protection – usually in the form of patents, but also covering internal proprietary *know-how* – represents a 'defensive' strategy. In this sense, inter-subjective trust is thus sanctioned by more formal and contractual arrangements that enable small life science firms to engage in relationships that would be characterised by severe disparities in power otherwise. Alongside this defensive strategy there is a concurrent need to collaborate with and actually trust other firms and organisations, which entails a clear negotiation over the incentives underpinning the involvement of both sides in any collaboration. In particular, the need for small life science firms to 'own' their knowledge means that they are unlikely to engage with other organisations that do not (or cannot) agree to such concerns.

The negotiation of distinct incentives and motivations behind knowledge production and, particularly, of IP ownership represents a good example of the complexities involved in the risk outsourcing in that it shows how governance strategies concern both the pursuit of value creation by small firms and the later value capture by multinational companies. The social processes that constitute this governance are embedded in inter-subjective forms of trust (e.g. competence, goodwill) that is necessarily underpinned by formal, contractual arrangements (e.g. patents). Overall this necessitates the development of closer, personal and direct relationships between individuals, which is less evident in the governance of knowledge relations discussed above; however, such inter-subjective trust is buttressed by more formal and contractual relationships (e.g. intellectual property rights), which provide a mechanism for engendering trust in relationships that might otherwise be characterised by severe power asymmetries.

## The Institutional Grounding of Industrial Transition

*Multiple Institutional Geographies*

In order to explore the importance of institutional geographies to successful upgrading in knowledge-based sectors, we want to highlight specific *institutional arrangements* that ground knowledge-based commodity chains in the Scottish life sciences. Here we look at public sector support, labour market dynamics and private finance in Scotland. We are interested in the particular forms that these institutions take, rather than the social norms or conventions that they engender (Pike et al. 2006) or the actors involved. Since this research was concerned with knowledge-based sectors and the (re-)positioning of firms along *global* commodity chains, we sought to avoid over-determining the 'regional' nature of these arrangements by considering the broader national and global institutional geographies and their influence on commodity chain governance (MacKinnon et al. 2002).

So, although Scotland's institutional arrangements (and environment) are unique, they also tied into broader national institutional arrangements. Research elsewhere illustrates the role played by such national institutional systems, especially in relation to less-favoured regions (LFRs) like Scotland (Birch et al. 2010). Just as important, however, are the global institutional arrangements, or more precisely the lack thereof (Birch 2011). Thus there is interplay between regional, national and global institutional arrangements, which needs to be acknowledged in order to understand the (re-)positioning of firms in knowledge-based commodity chains, especially in the life sciences, and the implications this has for industrial transition. The aim in this section is, therefore, to consider the specific institutional geographies of knowledge-based commodity chains in the Scottish life sciences and their implications for regional development.

*Public Sector Anchoring*

All the institutional arrangements discussed here are underpinned, in one way or another, by public support for a range of activities. One example is basic research funding, which has been increasingly oriented around a cross-university 'pooling initiative' in order to engender research excellence (Kitagawa 2009). In part this institutional arrangement is a response to the perceived physical and social geographies of Scotland. There is a perception that Scottish organisations, whether they were private or public sector and despite existing advantages, cannot compete individually in world markets. Therefore, collaboration is deemed necessary for future Scottish economic development.

This joined-up thinking is especially important when it comes to regional upgrading and broader processes of industrial transition as firms attempt to move up the commodity chain, which is necessary to capture value from the life sciences or, at least, stop it 'bleeding out' of Scotland. This contrasts, quite sharply, with the emphasis on individual entrepreneurship elsewhere in the existing literature on the life sciences. Such public sector support can be seen as a response to the Scotland's position in the global life sciences in at least two senses: first, Scotland is too small to represent a commercially-viable market; second, the Scottish life sciences industry is also small enough to engender trust across different organisations.

Alongside collaboration and pooling is an increasingly active engagement between the devolved government, universities and private sector firms in support of the life sciences. There are numerous examples of this engagement including the following:

- The *Scottish Seed, Co-investment* and *Venture* Funds set up by Scottish Enterprise that seek to bridge a number of financing gaps for new firms;
- The *Proof of Concept* programme also run by Scottish Enterprise that encourages technology commercialisation and transfer from universities;
- The 10-year funding for the life sciences through the *Intermediary Technology Institutes* (ITI) to encourage commercialisation of new science and technology; and
- The *Translational Medicine Research Collaboration* (TMRC) established by the large pharmaceutical firm Wyeth in collaboration with several Scottish universities and NHS Scotland.

These arrangements represent an alternative approach to regional policy-making in which public support is encouraged rather than viewed as 'crowding out' private investment (Birch and Cumbers 2007); i.e. a 'crowding-in' effect. However, what was also notable about such public sector anchoring of the life sciences was the deliberate global orientation of these decidedly regional initiatives. In particular, several of the schemes such as the ITIs and TMRC were designed or established to tie the Scottish life sciences into both global knowledge networks and global markets. Thus the public anchoring of the life sciences helps firms to (re-)position themselves in global commodity chains by not only providing financial resources but also strategic direction through support for collaboration, inter-organisational working, cross-sector coordination etc.

*Labour Market Dynamics*

Scotland's institutional arrangements are viewed as a model for successful upgrading and industrial transition elsewhere in the UK. Such perceptions, however, downplay the historical legacy of Scotland's strong basic science and clinical medicine base in its universities and teaching hospitals, which goes some way to explaining the wealth of university graduates. In one sense then, the labour market is particular to Scotland and when married to an autonomous National Health Service (NHS) it provides an industrial advantage few other parts of the UK have (or could replicate). This is particularly important for life science firms since the international scientific reputation – the consequence of local, ongoing public support – helps to tie Scotland into global knowledge networks. In this sense, scientific reputation ties Scottish firms, universities and so on into global knowledge networks and global commodity chains, helping to position these Scottish organisations at the frontier of research.

Without this, there would be little opportunity to attract and embed the global knowledge linkages. However, whilst the Scottish labour market is characterised by an abundance of new, highly-skilled entrants, experienced workers are much rarer. Firms are forced to recruit outside of Scotland, making them dependent upon the various global relationships they have. What is notable about this consequent need for firms (and other organisations) to seek senior staff from outwith Scotland (and often the UK) is that it is common for senior management positions to be filled by people returning to Scotland after they had left to gain commercial experience elsewhere. Thus there is an interesting labour market dynamic in which the emigration of people from Scotland caused by the lack of opportunities in Scotland itself then leads to the subsequent recruitment back into Scotland of those same individuals later in their lives. They thereby bring back valuable managerial capabilities that could not have been acquired otherwise.

*Private Financial Investment*

The development of these institutional arrangements goes some way to alleviate the perceived 'peripherality' of Scotland, which also features as an issue in the area of private finance and investment and which could also push Scottish firms out of knowledge-based commodity chains altogether. There is little institutional finance (e.g. pensions and mutual funds) or venture capital investment in the Scottish life sciences, but that as a consequence of this, and because of its peripherality, Scotland had developed a particularly strong business angel community.

In particular, the formation of business angels into syndicates provides another example of resource pooling which is the consequence of Scotland's local context. These syndicates combine individual business angel investments into larger funds, thereby creating larger and ongoing angel investments. The syndicates benefit from the various co-funding schemes set up by the Scottish Government (e.g. *Scottish Co-Investment Fund*). All this has led to the emergence of a very strong angel community in Scotland; this is not 'despite the fact that', but rather because there is limited (if no) institutional or venture funds available to life science firms. Consequently, Scottish life science firms have not been hamstrung by the lack of traditional financial investment sources; instead this supposed weakness has actually induced an important self-reinforcing mechanism.

Overall it is evident from the discussion of institutional embedding here that certain institutional arrangements underpin the (re-)positioning of Scottish life science firms in global commodity chains as part of a successful process of industrial upgrading that is in stark contrast to wider problems Scotland has faced with industrial transition. In particular, there is a strong public anchoring of the life sciences, especially in terms of funding for university research, start-up commercialisation, and early-stage firm growth. Such financial support is accompanied by a growing organisational support through the establishment of new agencies (e.g. ITIs), collaborative structures (e.g. TMRC), industry-government forums (e.g. Life Science Alliance), and strategic coordination (e.g. research pooling). These arrangements can be seen as a response to two over-riding imperatives. First, Scottish life science firms are, of necessity, 'born global' in that they have a limited 'home' market (e.g. Scotland, and even the UK) and are therefore always oriented to international markets, especially the USA. Second, the geographical, along with social and economic, peripherality of Scotland are seen as both disadvantageous and advantageous. On the one hand, the problems of peripherality are most obvious in the difficulties that Scottish life science firms have attracting certain forms of financial investment (e.g. venture capital), limiting their capacity to grow beyond a certain size in Scotland. However, on the other hand, this peripherality also helped to create a strong business angel community and an apparent greater capacity for collaboration between diverse partners.

## Conclusions

Our aim in this chapter has been to consider how the new forms of economic governance and institutional arrangements developing in knowledge-based commodity chains provide insight into regional upgrading and industrial transition in knowledge-based sectors. In this respect, there is considerable evidence, both in the life sciences and in other high technology industries, that relations of economic power differ from other less knowledge-intensive sectors because of the need to retain relatively open pipelines – both geographically and organisationally – to new and diverse ideas and knowledge (Bathelt et al. 2004). Multinational corporations find it less easy to dominate and control smaller firms and host regions than in most cost-driven sectors because the dynamics and uncertainties around knowledge production tend to work against the concentration and capturing of knowledge. At the same time, the realities of economic power and scale economies remain. Even in knowledge-based sectors, smaller innovative firms need the market reach and global distribution networks of multinational corporations.

We examined this underlying tension in the analysis of economic governance and institutional grounding of industrial transition, thereby illustrating the local-global dynamic in the Scottish life sciences. First, there are new forms of economic governance along the commodity chain. On the one hand, Scottish life science firms engage in numerous inter-organisational and extra-local relationships because they cannot integrate the necessary knowledge and capabilities in one organisation nor will these necessarily be contained in one place. These relationships are constituted by an objectification of trust in that adherence to international protocols and standards come to represent 'trustworthiness', which is then iteratively reproduced through the need for regulatory compliance along the whole commodity chain. On the other hand, firms have to ensure that in-house knowledge and expertise cannot be captured by other firms, especially multinational companies, or they lose their value and thus their ability to engage in collaboration. Such collaborative relationships are highly dependent on interpersonal forms of trust (e.g. competence and goodwill) that enable life science firms to align their interests (e.g. value capture through intellectual property protection) with those of the large multinationals that have the necessary global reach to market new technologies. In turn, large companies benefit from this arrangement because they are able to outsource the risk involved in research and development; namely the opportunity costs that long-term R&D necessitates.

Second, Scotland's institutional arrangements underpin this local-global dynamic. In particular, there is a strong, labour market, but a lack of managerial experience. However, although problematic, this is also beneficial because it means that Scottish firms have to attract skilled labour from outside of Scotland – often Scottish expatriates – who bring with them not only the requisite skills and experience but also a wealth of global connections. The same analysis applies to the lack of entrepreneurship and private investment. For example, whereas entrepreneurial drive and private sector profit-drive may be evident elsewhere, the anchoring role of the public sector proved particularly important for the Scottish life sciences. This not only includes financial support in the form of public funding, it also entails public-private partnerships, collaborative working and resource pooling. Finally, we argue that whilst the peripherality of Scotland – geographically as well as economically – may be seen as a weakness, it has meant that institutions in Scotland are underpinned by a close network of individuals, organisations and relationship which engenders high levels of trust because of the close social, physical and organisationally proximity with one another.

Nevertheless, with some exceptions, it is clear that the geography of knowledge-based commodity chains tends towards concentrating in particular regions and favours the more advanced regions and cities with the greatest assets of human knowledge, skills and financial capital. The life sciences sector is no different in this respect. Certainly opportunities exist for less-favoured regions to capture value and subsequent forms of economic growth, employment and new firm formation. However, such processes of industrial adaptation and transition are not an option open to all regions. Scotland's growth in the life sciences does itself reflect inherited institutional historical advantages from past processes of scientific knowledge production and the legacy of strong higher educational establishments plugged into broader global knowledge networks.


## Acknowledgements

This chapter is derived from two existing articles published in *Environment and Planning A* (Birch and Cumbers 2010) and *Growth and Change* (Birch 2011). The research for this chapter was funded by the Economic and Social Research Council (RES-000-22-2292). Special thanks go to Catherine McManus and Paul Jenkins for their invaluable research assistance, and to Mike Shand for producing the map. Obviously the views expressed here are those of the authors and the usual disclaimers apply.


## References

Amin, A. and Thrift, N. 1992. Neo-Marshallian nodes in global networks. *International Journal of Urban and Regional Research*, 16, 571–87.

Bair, J. 2005. Global Capitalism and Commodity Chains: Looking Back, Going Forward. *Competition and Change*, 9(2), 153–80.

Bathelt, H., Malmberg, A. and Maskell, P. 2004. Clusters and knowledge: Local buzz, global pipelines and the process of knowledge creation. *Progress in Human Geography*,28(1), 31–56.

Birch, K. 2008. Alliance-Driven Governance: Applying a Global Commodity Chains Approach to the UK Biotechnology Industry. *Economic Geography*, 84(1), 83–103.

Birch, K. 2009. The knowledge-space dynamic in the UK bioeconomy. *Area*, 41(3), 273–84.

Birch, K. 2011. 'Weakness' as 'strength' in the Scottish life sciences: Institutional embedding of knowledge-based commodity chains in a less-favoured region. *Growth and Change*, 42(1), 71–96.

Birch, K. and Cumbers, A. 2007. Public Sector Spending and the Scottish Economy: Crowding Out or Adding Value? *Scottish Affairs*, 58, 36–56.

Birch, K. and Cumbers, A. 2010. Knowledge, space and economic governance: The implications of knowledge-based commodity chains for less-favoured regions. *Environment and Planning A*, 42(11), 2581–601.

Birch, K., MacKinnon, D. and Cumbers, A. 2010. Old Industrial Regions in Europe: A Comparative Assessment of Economic Performance. *Regional Studies*, 44(1), 35–53.

Cooke, P., and Morgan, K. 1998. *The Associational Economy: Firms, Regions, and Innovation*. Oxford: Oxford University Press.

Cumbers, A., MacKinnon, D. and Chapman, K. 2003. Innovation, collaboration, and learning in regional clusters: a study of SMEs in the Aberdeen oil complex. *Environment and Planning A*, 35, 1689–706.

Gereffi, G. 1994. The Organization of Buyer-Driven Global Commodity Chains: How US Retailers Shape Overseas Production Networks, in *Commodity Chains and Global Capitalism*, edited by G. Gereffi and M. Korzeniewicz. London: Greenwood Press, 95–122.

Gereffi, G., Humphrey, J. and Sturgeon, T. 2005. The governance of global value chains. *Review of International Political Economy*, 12(1), 78–104.

Gertler, M. and Levitte, Y. 2005. Local Nodes in Global Networks: The Geography of Knowledge Flows in Biotechnology Innovation. *Industry and Innovation*, 12(4), 487–507.

Hopkins, T. and Wallerstein, I. 1986. Commodity Chains in the World-Economy Prior to 1800. *Review – Binghampton*, X(1),157–70.

Life Sciences Scotland. 2008. *Scottish Life Sciences Strategy 2008: Achieving Critical Mass*. Glasgow: Scottish Enterprise.

MacKinnon, D., Chapman, K. and Cumbers, A. 2004. Networking, trust and embeddedness amongst SMEs in the Aberdeen oil complex. *Entrepreneurship and Regional Development*, 16, 87–106.

MacKinnon, D., Cumbers, A. and Chapman, K. 2002. Learning, innovation and regional development: a critical appraisal of recent debates. *Progress in Human Geography*, 26(3), 293–311.

Massey, D. 1995, *Spatial Divisions of Labour*. 2nd Edition. Basingstoke: Macmillan.

Morgan, K. 1997. The Learning Region: Institutions, Innovation and Regional Renewal. *Regional Studies*, 31, 491–504.

Pike, A., Rodriguez-Pose, A. and Tomaney, J. 2006. *Local and Regional Development*. London: Routledge.

Storper, M. 1995. The resurgence of regional economies, ten years later: the region as a nexus of untraded interdependencies. *European Urban and Regional Studies*, 2, 191–222.

Storper, M. 1997. *The Regional World: Territorial Development in a Global Economy*. London and New York: Guilford Press.

Chapter 8

# Strategic Coupling between Multinational Subsidiaries and KIBS as a Mode of Regional Industrial Transition

Leo van Grunsven, Wouter Jacobs, Oedzge Atzema, Ton van Rietbergen

## Introduction

Globalization forces have prompted an increasing engagement of regions in developed countries with securing new sources of growth and of economic renewal through innovation, especially considering the volatility of internationalization and the ongoing processes of industrial shift (Buckley 2009). Leveraging multinational enterprises (MNE) and their subsidiaries is by now well acknowledged in the literature as a way for regions to accomplish economic development (Yeung 2009). Although there is still debate about the nature of regional spillovers of MNE subsidiaries (cf. Oetzel and Doh 2009, as introduction to special issue of *Journal of World Business* 44), there is increased awareness of the role multinational subsidiaries play in the industrial transition of regional economies (see Wang, Liu and Li 2009).

This makes their leveraging still opportune to these regions, as they perceive knowledge and innovation, embodied in regional enterprise and its networks, as the major sources of regional transition and strength. But regions have to confront the issue how to stay relevant, given a volatile competitive environment. Policy makers working on industrial transition are becoming aware that continued leveraging of multinational subsidiaries requires retention and a positive development trajectory in the host region. Thus, the questions have extended from simple location factors to the longer-run evolution of incoming subsidiaries of multinational corporations, their interactions with local suppliers and their contribution to regional development.

In thinking on regional innovation and economic transition, the role of Knowledge Intensive Business Services (KIBS) has gained scholarly interest since the early part of the last decade (Muller and Zenker 2001, Keeble and Nachum 2002). Research has made clear that – while within regional innovation policy the focus traditionally has been on manufacturing-related research and development – innovation activity increasingly engages with business services (Wood 2009). Innovations associated with KIBS are much more 'hidden' (because they are less technology-related and relying much more on informal relationships, cognitive proximities and customer-specific services), but in many advanced urban economies are considered to be crucial for international competitiveness in a knowledge-based economy (Wood 2009).

The literature on the evolution of subsidiaries of multinational corporations in the host region on the one hand, and the role of KIBS within regional innovation systems on the other hand remain somewhat distinct from each other. The objective of this chapter is to offer a conceptual understanding of regional industrial transition by combining the insights on multinational subsidiary evolution and on KIBS as a source of innovation. Recently, the notion of strategic coupling has been advanced by a number of scholars to elucidate how regions can engage with enterprise to promote transition and development (e.g. Yeung 2009). The second part of a conceptual understanding is arguing that *strategic coupling* of both sets of firms will lead to a process of co-evolution and can generate prospects for regional economic development. In doing so, we attempt to build stronger ties between insights from business economics with those of economic geography.

Our line of reasoning is as follows. Given an evolutionary path that increasingly emphasizes specialization of MNE-subsidiaries on knowledge-intensive tasks, we suggest that KIBS can be considered as a major asset of the local host environment of these subsidiaries. KIBS can contribute to learning processes of the MNE subsidiary and, henceforth, to the subsidiary's further evolution within the overall corporate network. Likewise, KIBS are able to thrive through the presence of multinational subsidiaries that not only constitute a source of business, but also provide them the opportunity to learn from these client-interfaces, possibly leading to specialization. Thus, a source of regional transition lies in interactions between multinational subsidiaries and KIBS whereby each is an important constituent part of the local external environment of the other. Stated otherwise: the presence of the one in the region constitutes a 'relational asset' to the other one in terms of learning, innovation and upgrading. In terms of industrial transition and regional development it is then the strategic coupling of both that provide best prospects for success.

This contribution is therefore structured as follows. In the next two sections, we review the literature within business economics on MNE subsidiary evolution and highlight the increased importance given to the local host environment as an explaining variable in the evolution of subsidiaries. In particular we want to address here the role of KIBS as a relational asset of MNE-subsidiaries within the host environment. In the fourth section we invoke the concept of 'strategic coupling' as a means to stimulate the relational interactions between MNE-subsidiaries (as part of global networks) and local-based KIBS suppliers in order to promote regional economic development. In the fifth section, we address our conceptual framework as well as important contingencies that need to be taken into account.

## Multinational Subsidiaries, their Evolution and Regional Development

Subsidiaries – defined as 'a semi-autonomous entity with entrepreneurial potential, within a complex competitive arena, consisting of an internal environment of other subsidiaries (... internal customers and suppliers), and an external environment consisting of customers, suppliers and competitors' (Birkinshaw et al. 2005: 227) differ in terms of activities they carry out, in their relation to other subsidiaries and their headquarters and in their own strategic responsibilities. The literature (for example, Birkinshaw 1996, Delany 2000, Dörrenbächer and Gammelgaard 2006, Paterson and Brock 2002) refers to subsidiary mandates. These, and especially their evolution, are significant from a regional perspective.

Mandates generally refer to 'the businesses, or elements of a business, in which the subsidiary participates and for which it has responsibilities beyond its national market' (Birkinshaw 1996: 471). Mandates can range from a small part of the product chain, for example only sales, to total responsibility of a product in which the subsidiary is given 'global responsibility for a single product line, including development, manufacturing and marketing' (Birkinshaw 1996: 468). Activities carried out by subsidiaries can comprise different mandates (Delany 2000).

Mandates are reflected in the market, product and value-adding scope of activities (White and Poynter 1984, in Dörrenbacher and Gammelgaard 2006; see also Wang et al. 2009 for a review). Market scope pertains to the geographic markets served; product scope relates to product line variety, extension or new product areas in a subsidiary's business; and value-adding scope relates to the number of stages in the value-added chain, performed in the host economy. Underlying a specific scope is the stock of competences that are present in the subsidiary. While scope and competences are interrelated, the relevant aspects here are their evolution and the sources of competences (development) in the framework of the entire range of forces that may impinge on profile and evolution.

Evolution of activity mandates of MNE subsidiaries reflects changes along one or more of the dimensions of market, product or value-adding scope. While it is argued that changes in scope – often related to both organisational and spatial considerations – tend towards broadening (for example, Benito et al. 2003), the opposite is possible also resulting in mandate loss or removal (for example, Dörrenbacher and Gammelgaard 2010). Bundling, unbundling and rebundling of activities occur frequently, as the number of subsidiaries is rationalized (associated with for instance merger or acquisition activities) or when local environmental conditions prompt change of responsibilities (Birkinshaw 1996, Benito et al. 2003, Dörrenbächer and Gammelgaard 2006).

Subsidiary development implies also a process of bundling, unbundling or rebundling of competences. Again, this can go in different directions: an increase of the (scope of) activities requires a larger and more diffentiated set of competences; in case of un- or re-bundling of activities competences may be sacrificed, or transferred (Benito et al. 2003). Some researchers (for example, Birkinshaw 1996) view subsidiary capabilities solely as a determinant of subsidiary development.

However, a 'desired' scope of activities may also be a driver of competences acquisition through a range of mechanisms.

Work on evolution of mandates interprets re-production or reconfiguration as the preservation and renewal, abandonment, substitution or addition of functional parts (Dörrenbacher and Geppert 2009). Mandate evolution can produce either growth or shrinkage and takes place in two ways. First, quantitative in terms of change in scale and geographical scope: a subsidiary carries out an activity and extends or downsizes this activity over time, both with respect to size and intensity. An example is a sales activity developing into European sales activities. Second, qualitative in terms of adaptation of responsibilities. This ranges from diversification, broadening or narrowing, substitution, and deepening. Such qualitative change impacts on the position of the subsidiary in the enterprise.

In management and international business studies subsidiary change is often seen to be driven by change over time in the organisation of the MNE, distinguishing several internal and external determinants. The – evolution of – the mandate(s) of a subsidiary are seen to be determined by (1) the strategies of headquarters and the other subsidiaries, (2) the subsidiary choices (both in part reflecting – dynamics of – the international and national environment) and (3) the local external business environment (Birkinshaw and Hood 1998, Dörrenbächer and Gammelgaard 2006; Madhok and Liu 2006; Paterson and Brock 2002). Below we will comment on this interpretation, whereby we suggest that some of the 'determinants' are often interpreted too narrowly. We point to alternative interpretation and we will argue that some of the above-mentioned factors should be considered as contingencies in an explanatory framework rather than determinants.

## Local Host Environment as a Relational Arena

As indicated above, part of the interpretation of firm dynamics in regions relates to the local external environment. From the perspective of multinational subsidiaries however there is reason to quarrel with the usual conception of the host-environment and the relationship between host-environment and subsidiary evolution. In the literature the host-environment is usually seen as a set of conditions, mostly at macro-level (for example, Birkinshaw, Hood and Young 2005). They can be availed of by subsidiaries, and often are as they impact monetary return and profitability. While these factors constitute comparative (dis-)advantages, such an interpretation of (the role of) the local host environment has been challenged as being limited by focusing on static generic aspects. A micro-oriented conception that considers specific aspects in a dynamic fashion from the perspective of the subsidiary has been the point of departure of work that has developed a broader understanding of the local external environment. There is no reason to question the validity of the focus of Birkinshaw and Hood (1998) on the subsidiary's accumulation (or depletion) of capabilities over time as a driver of subsidiary development, as well as to the understanding of these capabilities as being special relationships, business-specific competences, growth-enabling competences and privileged assets (Delany 2000). At the same time there have been a number of attempts to deal with the host environment as the place where resources are situated that enable accumulation of specific rather than generic assets, thus allowing the bundling of competences. This goes some way in coming to grips with the mechanisms underlying the building and maintenance of competences of a subsidiary. Insourcing networks and collaborations, whereby special relationships with diverse sets of actors develop, suggest a conception of the host environment as a *relational arena*, and the relationship between subsidiary and host environment an active one (Schmid and Schurig 2003). This relationship has the capacity to increase autonomy for the subsidiary and provide them insight in how to add value. This is a relational approach that sets subsidiaries as (central) actors in an evolving host-environment-situated network. In this way a *relational asset* conceptualization of the relationship between MNE subsidiaries and their environment is being developed.

Some elaboration can be found in the recent literature. Particularly useful for our purpose is the contribution by Dörrenbacher and Gammelgaard (2010), building on insights developed by Schmid and Schurig (2003) and Wang et al. (2009). The former have demonstrated how capabilities that are performed by the subsidiary better than other units within the same enterprise are developed within the foreign subsidiary from its network of relationships. A large part of these relationships concerns those with external actors in geographical proximity to the subsidiary, who are thus constitutive of the local external environment. One view in this domain is that local networks serve mainly to absorb resources relevant to the enterprise as a whole (often with local transformation into competences) whereby subsidiaries serve as transmitters: nodes in a network of global pipelines developed by the enterprise. The subsidiary stays as long as it can usefully fulfil this function. Another view however emphasizes own initiative by subsidiaries to utilize these resources to enhance or transform its competences allowing a deepening or transformation of the market-, product- and/or value added scope (that is through specialization). In this way they can stay relevant to the enterprise by generating something 'new'. It should be noted that this approach has a dynamic element only in hypothesizing that relationships will change from internal to external with age of the subsidiary.

Wang et al. (2009) suggest that a 'product-market-value added'- framework – while valid – misses out on the mechanisms by not going into interactions between subsidiary and local environment. In their view, a network approach is best suited to conceptualize the local environment as the crux of interaction with local external business partners which provide access to resources. They qualify that there is differential behavior in this respect, producing different subsidiary performance. More interaction induces better performance, translating in a better overall position in the larger enterprise.

Dörrenbacher and Gammelgaard (2010) relate changes in charter or task-related activities to inter-organizational networks, both within the enterprise and in the local environment. Interaction may either produce new subsidiary resources and competences or, in a negative way, lead to their demise and render competences to irrelevance. Localized inter-organizational networks is about creating linkages with critical suppliers, as well as mobilizing institutional resources through engagement with government agencies. This is concretized through the intra-enterprise networks, mediated by the headquarters' objectives and strategies. The authors in this context invoke the notions of network infusion and density and network centrality. The former concerns the frequency and depth of linkages, shaping but also deriving from subsidiary entrepreneurship. The latter is relative to headquarter and other units in the enterprise, and reflects the value of localized networking to other parts of the enterprise through transmission of resources and competences. Subsidiary initiative plays a role in shaping this value. A subsidiary may be relegated to the periphery of the enterprise in case of: 1) infrequent linkages to the parent or head office; 2) poor performance of the subsidiary in terms of value added to the enterprise, or 3) the case in which the head office decides to set priorities on markets or business segments different from the ones in which the subsidiary is active. The above illustrates precursors and concrete instances to the idea of the local external environment as a relational arena from the perspective of multinational subsidiaries.

*KIBS as a Relational Asset of the Local Host Environment*

Within business economics, the importance of local relationships has thus been recognized as an important factor for the performance and evolution of MNE-subsidiaries. However, much of the emphasis thus far has been put on local-based suppliers of production inputs, on professional organisations such as universities and on regulators, distributors, customers and competitors (Schmid and Schurig 2003). What has been largely ignored is the role of so-called knowledge intensive business services or KIBS as a critical asset of the local host environment. Within economic geography, in particular those studies that focus on regional innovation systems (RIS), there also tend to be a conceptual neglect of KIBS (but see Muller and Zenker 2001). Despite the intellectual influence of the concept of RIS within academia and especially within policymaking over the last decade, it tends to be biased towards manufacturing-related research and development activities that have a focus on the development of new technologies and of new products that are both based upon 'hard science'. While we do not want to downplay the importance manufacturing related research and development, we do want to emphasize the conceptual neglect of the role of KIBS. This neglect is problematic because in many advanced industrial economies services tend to be the main source of employment and economic growth (OECD 2005). Service innovation, especially in business services, are much more 'hidden' because they are less traceable (e.g. through patents) and more tacit in nature. They rely much more on close interaction, on cognitive proximities and on customer-specific, intangible products. Within the knowledge-based economy, the presence of KIBS can then be considered as crucial for the international competitiveness of cities and regions (Simmie and Strambach 2006).

There is not one standard or universal definition of what constitutes a KIBS. Here we make use of the definition provided by Den Hertog (2000: 505): 'private companies or organizations that rely heavily on professional knowledge, i.e., knowledge or expertise related to a specific (technical) discipline or (technical) functional-domain to supply intermediate products and services that are knowledge based'. In essence, KIBS provide services such as local market information to foreign investors, business models to manage complex corporate functions across space (e.g. supply chain management), customized ICT-solutions; all of which help their clients to deal with uncertainties in the (global) market economy.

According to Strambach (2008: 156) three main features of KIBS can be discerned that, despite their heterogeneity, can be considered as their defining commonalities. The first feature is knowledge as product. Knowledge is not only the key production factor of KIBS, but also the ultimate product they sell. KIBS largely provide and sell non-material intangible services by making use of a high educated and skilled workforce. The second feature is in depth interaction. KIBS produce their services through intense interaction with their customers in which both parties are involved in cumulative learning processes. This interaction results in customized rather than standardized solutions and services. The third feature is activity of consulting. KIBS activity involves a process of problem solving in which they can adapt their expertise and expert knowledge to the specific needs of the client.

Strambach (2008) continues to further categorize KIBS by means of the type of knowledge they use and produce. She distinguishes three types of knowledge and organizes different types of KIBS accordingly. The first type of knowledge is analytical knowledge, which tends to dominate in industries in which science-based knowledge is important. The knowledge creation process in this category tends to be highly formalized into step-by-step rational research procedures in which codified science is applied and tested. In general, this kind of knowledge creation is more appropriate within manufacturing related research and development (e.g. chemicals) and not so much in KIBS. Nonetheless, specialized research and development service firms form a small but growing group that do contract related research for often large international clients (Strambach 2008: 158). The second type of knowledge is synthetic knowledge. Characteristic for this type of knowledge is that innovation often takes place in the form of new combinations of knowledge. This type of knowledge base is more typical for KIBS, in particular for traditional professional services such as (management) consultants. In general, synthetic knowledge relies much more upon tacit knowledge acquired through intense interaction with the customer and in which customer's problems are solved through past experience and learning by doing. Knowledge derived from past experience will be recombined with the customer's specific problem resulting into customized services and solutions. A third category of knowledge base is symbolic knowledge. Symbolic knowledge is much related with creative or cultural industries which deal with external communication, symbols, logos and design. It depends on culturally defined interpretation schemas and other informal conventions such as habits, norms and taboos. With regards to KIBS, this type of knowledge is particular relevant for marketing and advertisement.

Important to realize is that at a sectoral level KIBS operate within complex horizontal and vertical knowledge domains in which their own knowledge bases are located (cf. Strambach 2008). Horizontal knowledge refers to certain business functions, whereas the vertical knowledge domain refers to sector-specific specialization. Different business functions within corporations are typically discerned such as management & administration, research and development, marketing & sales, production and distribution and so on. These intra-firm business functions are most likely to be spatially dispersed among the corporation's different establishments. For all these business functions there is a need for particular type of KIBS. The vertical disintegration of production systems has been going on for a long time. This is certainly true for certain manufacturing processes of large industries (e.g. automotive) in which flagship companies vertically specialize in research and development and marketing functions, while increasingly outsourcing production processes and material supply to external partners. The same is however true for more business service related functions. Human resource management, legal services, ICT, executive management training and the like have increasingly become obtained from external, specialized KIBS. The result of these processes is that certain KIBS do not only need to possess certain horizontal knowledge with regards their clients' business functions, e.g. marketing services supplied to the marketing and sales department of their client, but also that they can or need to specialize in certain sectors or industries.

In addition to the observation that KIBS act as vectors of information exchange, economic geographical scholars started to analyze the locational behavior of KIBS. These studies suggest that KIBS cluster in large metropolitan areas due to agglomeration benefits they enjoy with their (global operating) clients (Keeble and Nachum 2002, Wood 2009, Muller and Doloreux 2009). Co-location of KIBS and their MNE clients is based upon the presence of positive externalities, most notably increasing returns to scale, knowledge spillovers and a specialized labor force. More specifically, given the tacitness of knowledge involved in KIBS-activity and given the need for intense interaction with their clients in order to develop such knowledge, learning and innovation is fostered by spatial proximity.

Thus, KIBS act as creators, carriers and diffusers of knowledge through their client-interfaces and they tend to co-locate nearby each other and nearby clients. As such they can be perceived as a relational asset for MNEs within a particular local host environment. But KIBS do not only provide customized knowledge to MNE clients; they also learn themselves through these client interfaces. This generates the possibility of a process of co-evolution between both populations.

## Co-Evolution Derived from Strategic Coupling between MNEs and KIBS

In building our conceptualization, first of all we invoke a economic geographic concept that recently has been popularized in regional development studies and relational economic geography, namely strategic coupling. Second, we also invoke a concept from evolutionary economics (or evolutionary economic geography), namely co-evolution. Application of these concepts allows a more stringent formulation of environmental relationships.

The concept of 'strategic coupling' has been coined by scholars involved in the study of global production networks and regional development (for a review, see Yeung 2009). Uncomfortable with the endogeneity perspective, these scholars recognize the extra-regional processes that shape regional trajectories in the era of contemporary globalization, whereby global networks of firms act as pipelines through which new knowledge, capital, instructions and talent is pumped into the regional economy (Bathelt et al. 2004). Attracting these globalized firms becomes crucial for regional economic development. Strategic coupling is defined as 'a contingent convergence of interests and cooperation between two or more groups of actors who otherwise might not act in tandem for a common strategic objective' (Yeung 2009: 332). Coupling is strategic in the sense that it is purposive intervention by one of the participants. This can be a government agency or state actor concerned with regional economic development, but might as well be a (coalition of) firms searching for new business opportunities. While there is a convergence of interests of various actors, these by no means are restricted to the same spatial scale. Some have stakes at geographically dispersed sites whereas others are only limited to localized interests. The object of strategic coupling revolves around how resources are mobilized through relational interdependencies between localized actors and global players.

*Why Multinational Subsidiaries Couple with KIBS, and KIBS with Multinational Subsidiaries*

We apply the concept here to understand the interaction between multinational subsidiaries and KIBS within a regional context (as an added driver to the process of regional industrial transition, besides these types of firms in isolation). Strategic in this context refers to an intentional linkage-process in which specific demands of one set for particular resources become successfully matched with local supply by another set. This needs clarification of the object and purpose of the coupling process from the specific perspective of these types of firms.

The local presence of KIBS can become a relational asset for foreign investment and multinational subsidiary evolution in the dimensions discussed earlier. In identifying the imperatives driving strategic coupling, it should be noted first that value chain strategies of MNEs in the locational aspect at the current juncture in mature regions emphasize fine slicing and selective co-location (Defever 2006). What can be observed is increasing functional *specialization* of subsidiaries.

Concurrently, in value chain strategies new routines have been developing that substitute at least in part in-house production or execution of all steps related to tasks by localized insourcing.

For subsidiaries, this implies that increasingly, a transition towards higher order mandates is specialization driven. In this context, our own empirical research reveals that in the EU context specialization increasingly is on *knowledge and information-intensive* tasks (Gerwen van and van Grunsven 2010). Thus, on the one hand the role of KIBS reflects such strategies and path of evolution of subsidiaries. On the other hand, the accumulation of competences driving patterns of specialization and co-location is assisted by the presence of KIBS. Subsidiaries derive power from performing well in a specialized mandate across products and/ or being 'on the radar' as to internal competition for the co-location of functions that increase value added. Such a performance and intent provides a significant incentive to link with KIBS.

The latter follows from the fact that at the current juncture – as subsidiaries in mature regions incrementally specialize on knowledge- and information-intensive tasks – knowledge and information is becoming the most privileged specific regional asset. Traditionally, the emphasis as to direction of knowledge spillovers has been the host environment's knowledge stock changing due to the presence of the subsidiary, which in turn has received knowledge from the headquarters. However, some authors, such as Benito et al. (2003), have started to acknowledge the influence of the environment on the development of subsidiaries in terms of exchange of resources focused on knowledge, codified and tacit. Bathelt et al. (2004) state that knowledge transfer and attainment elsewhere only become valuable if combined with knowledge embedded *in a local environment* in tacit forms. This prompts a conceptualization of – the dynamics of – the local environment in terms of knowledge production and consumption, rather than as a repository whereby knowledge is considered to be simply there, ready to be absorbed. The relevant mechanism is the *active and specific production in networks* between subsidiary firms and localized knowledge sources, public and private. The creation and accumulation of knowledge within subsidiaries increasingly occurs through their interactions with local firms. The added relevance to subsidiary evolution is that knowledge partially determines a subsidiary's autonomy (see also Madhok and Liu 2006). We perceive that increasingly the stock of relevant local knowledge is 'embodied' in localized KIBS as these are able to receive information from outside the firm, i.e. from their host environment, and to combine this information with firm-specific knowledge resulting in useful, increasingly indispensable, services for customers. These KIBS are not only locally based suppliers. Research on world cities (e.g., Taylor 2004) continues to emphasize how leading business services agglomerate within certain places and in proximity of their MNE clients, specifically to facilitate their clients' global business within spatial proximity.

The knowledge provided by KIBS includes a high degree of tacit knowledge, so KIBS contribute to the tacit knowledge base of their clients. However, it is not true that KIBS are only knowledge suppliers and their clients are the receivers.

The knowledge they possess results from a co-production process involving their clients, constituting a two-way process (Muller and Doloreux 2009). The local presence of MNEs does not always provide KIBS with sufficient market demand. However, it also allows for KIBS to become engaged with the MNE's global network. Indeed, the world city research referred to above has highlighted how high order producer services or KIBS have both facilitated and followed the internationalization strategies of their clients. At the local level, however, continuous interaction with MNEs also provide KIBS with more complex and demanding projects as well as with increased reputation. Continuous local interaction with MNEs therefore allows KIBS to develop skills and possibly to specialize in the specific sector or industry of the MNE or in specific types of business functions. In other words, KIBS can develop both horizontal knowledge domains in terms of specialized business functions of their international clients (e.g. marketing & sales), as well as vertical knowledge domains such as the marketing of Japanese consumer electronics products in local markets (Strambach 2008).

*Co-Evolution*

The mutual relations between development of, and innovations in, firms and the environment can be interpreted at the micro-level as a kind of co-evolution with firms in a localized web that tends to become denser over time. Madhok and Liu (2006) however postulate 'multi-levelness' and 'hierarchical nestedness" (Madhok and Liu 2006: 3) as important notions when using a co-evolutionary perspective, arguing that co-evolution occurs at multiple separate levels, with the units of evolution being nested within one another. The following quote explains the micro-level: 'the emphasis is on the simultaneous evolution of organizations […] and their environments – comprising other organizations and entities – where the former influences the latter as well as vice versa in a continual and interactive process' (Madhok and Liu 2006: 3).

A macro co-evolutionary perspective is possible from a broader definition, namely 'a situation in which two or more evolutionary systems are linked together in such a way that each influences the evolutionary trajectory of the other(s)' (Safarzynska 2009: 42). These evolutionary systems can be sets of firms and organizations and, at a higher aggregated level, regional economies and institutions. According to Lee and Saxenian (2008), a critical question concerning the process of co-evolution, understood at various analytical scales, becomes: 'how to make theoretical connections between micro events that happen at the firm level, and their spatial repercussions that can usually be observed only at the regional level' (Lee and Saxenian 2008: 158). An often argued indicator of co-evolution regionally is agglomeration and co-location, enabling spatially concentrated firms to benefit from positive externalities and increasing returns (Boschma and Frenken 2010). The reasoning is that knowledge spillovers, learning and innovation are fostered by spatial proximity, promoting the adoption of new organizational routines or new technologies. Moreover, interaction between firms in geographical proximity can result in spinoffs, while a large qualified localized labor pool generated by firm agglomeration generates startups in combination with market potential. Spatial concentration is then enhanced, as spinoffs are more likely to stay close to the mother, while startups are more likely to emerge close to the entrepreneurs' home and to their social networks.

## Strategic Coupling, Co-Evolution and Regional Industrial Transition

In our conceptualization we elaborate on the relationship between firm dynamics (both multinational subsidiaries and KIBS) and the host environment (of which both types of firms are a constituent part) through the lens of strategic coupling. Our position is: we can understand firm evolution and the role of the local environment as a process of strategic coupling and co-evolution; the coupling of knowledge-intensive business services and multi-national enterprises in time and space render each a relational asset to the other exploited at firm-level in competence development and activities portfolio, at industry-level in industrial specialization patterns and at regional level in industrial transition.

Multinational subsidiaries and KIBS enter into regional cooperative arrangements (e.g. innovation) in order to take advantage of each other's complementary assets. This continuous relational interaction generates new sets of *relational assets* which at the micro-level are exploited to develop competences and new routines allowing product-market-value added upgrading. Assets thus are continuously produced, re-produced and re-configured. Regions that become endowed with sufficient relational assets, when successfully mobilized, possess a competitive advantage over others and promote sets of firms to evolve by the development of capabilities and routines, also enabling deeper specialization in knowledge-intensive activities in a specialized segment of the firm product portfolio. Co-evolution whereby firms exploit room for initiative-taking and autonomy produces on the one hand outcomes that influence their power position (centrality) vis à vis other units in the enterprise and headquarters, in turn influencing survival and development path. On the other hand the embedding of firms into the region in this way also produces unique and intangible or inimitable resources for the regional economy, in the form of spinoffs and start-ups.

The conceptual relations between multinational subsidiaries and KIBS are summarized in Figure 8.1, in which we distinguish two different trajectories (a and b) both starting at t1 with the presence of a MNE-subsidiary within a particular local host environment. At t1a, the MNE-subsidiary starts to engage with KIBS in the local host environment (relationship 1), directly and independently from the MNE head office. Such interaction takes place formally through transactions, contractual agreements and joint projects and informally through all kinds of social networks and local business circuits. Continuous interactions in close (spatial) proximity can lead to positive agglomeration externalities (knowledge spill-overs), and whereby transfer of tacit skills and 'embodied knowledge' can

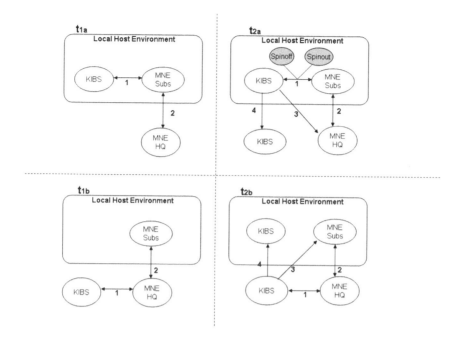

**Figure 8.1    Conceptual relations between MNEs and KIBS in the host environment**

also occur in the form of labour mobility. These generate prospects for routine breaks within the MNE-subsidiary and in which the MNE-subsidiary develops more knowledge-intensive activity. The result of which might be improved centrality within the MNE-corporate network, or at least the assignment of more competences or improved mandates by the head office (relationship 2).

Indeed, acquiring competences and functional mandates through strategic partnerships with local based KIBS (*relationship 1*) is a relevant strategy of the MNE-subsidiary to strengthen its position within the overall MNE-corporate structure (relationship 2; cf. Delany 2000, Paterson and Brock 2002). At the same time, KIBS benefit from these interactions through improved reputation or, more importantly, by means of learning processes that might result in specialization in certain vertical knowledge domains, that is certain industrial sectors (cf. Strambach 2008). At t2a, the following can have happened. Joint project work and labor mobility between KIBS and MNEs can produce spinoffs, spinouts and startup companies in the local host environment. Spinoffs and spinouts are indeed more likely to locate nearby the parent company, whereas startups are likely locate nearby the entrepreneurs home. Another option is that through the local interactions with the MNE-subsidiary, KIBS gain access to the MNE's global network and built up expertise and reputation. KIBS can then provide services directly to the MNE headquarters (relationship 3) or, more likely, they are stimulated to internationalize themselves and set up a corresponding international network to service their global operating clients (relationship 4).

We should however also consider a different trajectory, one in which the MNE head office engages with KIBS outside the local host environment of the subsidiary at t1b. Based upon this relationship, KIBS are hired centrally by the MNE head office to provide services to the subsidiary directly at t2b (relationship 3) or to set up a local division in order to service the subsidiary locally (relationship 4). For the MNE that operates globally, but under numerous local conditions and contexts, there will remain incomplete knowledge about local (market) conditions. MNE subsidiaries will provide feedback to the MNE head office about the instructions, manuals and applications received, taking into account these local conditions. For this same reason, KIBS are therefore inclined to set up local offices as well, often by taking over established local firms or by setting up local divisions under the management of a local-based partner. Once established within the local host environment, KIBS can service the MNE-subsidiary directly with possibility of generating spinoffs, startups and spinout similar to relationship 1 at t2a.

In both trajectories, strategic coupling between both set of firms generates prospects for regional development and industrial transition. However, we should be aware that the model presented is somewhat stylized and that the empirical reality is much more messy than the model suggest. We should in particular be aware of certain contingencies that shape the evolutionary pathways as addressed.

*Contingency: Selectivity and Regionally Specific Pathways*

Although it is argued that some of the relationships that are conducive of learning and innovation have been de-territorialized, and non-local relationships in the form of global pipelines pump knowledge, technology and business practices from elsewhere into regions, for reasons discussed in the part on co-evolution, spatial proximity of KIBS often still is a necessary condition for subsidiaries to make use of KIBS knowledge and services (Muller and Doloreux 2009), and the other way around. Co-evolution on a local basis is relevant, shaped by unique historical, institutional and strategic conditions that influence outcomes in time and space. There is apparently a positive mechanism taking place in which both KIBS and their international operating clients benefit from each others' spatial proximity. Co-location therefore is seen as a good first indication of coupling. However it is not necessarily evidence of it; that is, not necessarily a sufficient condition. In this context, Boschma and Frenken (2010) allude to a – potential – spatial proximity paradox that can arise from a range of factors. One factor they mention is the need for cognitive proximity in order for actors to build absorptive capacity and understand each other. Cognitive proximities, trust and a shared labour pool of related human capital facilitate the learning process among international orientated businesses and local-based KIBS.

It should be recognized that there are other factors or conditions as well, of a differentiating nature, likely circumscribing the operation of the stylized model outlined in the previous sub-section. Co-evolution between firms through continuous interaction in various arenas does not take place in a continuous or uniform manner. Three 'categories' of 'contingencies' may be distinguished: general, enterprise/headquarter-related, and individual firm/establishment-related. We conclude this part by highlighting a few selected aspects within each of these categories. As to the general category, those who have advanced the strategic coupling 'theory' have added the qualification that coupling is time-space contingent (Yeung 2009). It is not permanent and subject to change as it resembles a temporary coalition of actors and institutional arrangements. Also, heterogeneity or variety of the population probably is an important aspect to consider as the above trends likely occur and work out differentially in different sectors (manufacturing and services) and branches. Next, the absorptive capacity of actors to learn from interaction is also determined by past experience in their environment. This implies that regional and local industrial legacies (and established institutional frameworks), also can act as constraints by excluding novelties that do not correspond well with the existing knowledge base, vested interests and existing institutional architecture.

As to the enterprise/headquarter-related, and individual firm/establishment-related categories, one idea derived from the literature (*inter alia* Birkinshaw and Hood 1998, Buckley 2009, Delany 2000, Dörrenbächer and Gammelgaard 2006, Dörrenbächer and Geppert 2009, Paterson and Brock 2002, Schmid and Schurig 2003) is that enterprises and their headquarters (HQs) employ widely divergent corporate strategies (heterarchical vs. hierarchical; multi-domestic vs. global; realised vs. HQs' intended), translating into differential HQ influence on the development of subsidiaries. Another idea is that subsidiaries – as a reflection of origin – are imbued with organizational routines employed by corporate management derived from differential cultural settings. As subsidiaries inherit these culturally determined organizational routines through transfer top down, some are more able to build local strategic partnerships than others. Next, the – initial – mandate and function(s) of subsidiaries matter. Function(s) translate(s) into differential local need for KIBS and capacity of the establishment to strategically engage with KIBS (compare for instance sales and research and development). Subsidiaries still differ in terms of their initiative taking and entrepreneurial efforts, in the effort by the management to increase autonomy and develop its own network, although these are seen to help in legitimizing the existence of the subsidiary. Finally, Institutional settings differ between regions and may either enable or constrain the strategic coupling between local based KIBS and MNE subsidiaries.


## Conclusions

In this chapter we have dealt with regional industrial transition from the perspective of the interaction between two sets of firms: subsidiaries of multinational corporations and KIBS. As knowledge creation and diffusion has taken centre-stage in the upgrading of MNE-subsidiaries, we argue that strategic engagement with KIBS emerges as a relational asset. Local based KIBS in turn will become connected with internationally operating clients and their specific demands for knowledge. Such inter-dependency generates prospects of co-evolution. Core of the argument is that strategic coupling between MNE subsidiaries and KIBS implies a significant channel for innovation and regional industrial transition. We have primarily been engaged with conceptual development through deductive reasoning, by making inferences from existing research and literature in the field of business economics and of economic geography. While economic geographers are moving into the territory of business economics by explaining regional economic development outcomes in terms of the evolution of organizational routines at the micro-level, business economists are increasingly aware of the impact of the local host environment on the performance and evolution of business units within the corporate network. We have argued that combining the insights from both disciplines can yield further conceptual understanding of the ongoing processes of regional industrial transition taking place in the world today. In particular we have addressed the role of KIBS in these processes, understanding them not purely as a tertiary sector in itself, but as a crucial local provider of knowledge intensive services to their (manufacturing-based) clients. Further empirical case study work, taking into account the contingencies addressed, is needed in order to assess the co-evolutionary processes and to induce hypotheses for further theoretical inquiry.

## References

Bathelt, H. and Glückler, J. 2003. Towards a relational economic geography. *Journal of Economic Geography,* 3, 117–44.

Bathelt, H., Malmberg, A. and Maskell, P. 2004. Clusters and knowledge: local buzz and global pipelines and the process of knowledge creation. *Progress in Human Geography*, 28, 31–56.

Benito, G. R. G., Grøgaard, B. and Narula R. 2003. Environmental influences on MNE subsidiary roles: economic integration and the Nordic countries. *Journal of International Business Studies*, 34(5), 443–56.

Birkinshaw, J. 1996. How multinational subsidiary mandates are gained and lost. *Journal of International Business Studies*, 27(3), 467–95.

Birkinshaw, J. and Hood, N. 1998. Multinational subsidiary evolution: capability and charter change in foreign-owned subsidiary companies. *Academy of Management Review*, 23(4), 773–95.

Birkinshaw, J., Hood, N. and Young, S. 2005. Subsidiary entrepreneurship, internal and external competitive forces, and subsidiary performance. *International Business Review*, 14, 227–48.

Boschma, R.A. and Frenken, K. 2010. The spatial evolution of innovation networks: a proximity perspective, in *Handbook on Evolutionary Economic*

*Geography*, edited by R. Boschma and R. Martin. Cheltenham: Edward Elgar, 120–35.

Buckley, P.J. 2009. Internalisation thinking: from the multinational enterprise to the global factory. *International Business Review*, 18, 224–35.

Defever, F. 2006. Functional fragmentation and the location of multinational firms in the enlarged Europe. *Regional Science and Urban Economics*, 36, 658–77.

Delany, E. 2000. Strategic development of the multinational subsidiary through subsidiary initiative-taking. *Long Range Planning*, 33, 220–44.

Dörrenbächer, C. and Gammelgaard, J. 2006. Subsidiary role development: the effect of micro-political headquarters-subsidiary negotiations on the product, market and value-added scope of foreign-owned subsidiaries. *Journal of International Management*, 12, 266–83.

Dörrenbächer, C. and Gammelgaard, J. 2010. Multinational corporations, inter-organizational networks and subsidiary charter removals. *Journal of World Business*, 45, 206–16.

Dörrenbächer, C. and Geppert, M. 2009. A micro-political perspective on subsidiary initiative-taking: evidence from German-owned subsidiaries in France. *European Management Journal*, 27, 100–12.

Den Hertog, P.D. 2000. Knowledge-intensive business services as co-producers of innovation. *International Journal of Innovation and Management*, 4 (4), 491–528.

Gerwen, B. van and van Grunsven, L. 2010. *Overleven ondanks Padafhankelijkheid (Survival despite Path Dependency). Kwalitatief Onderzoek naar de Evolutie van Lang-Aanwezige (Buitenlandse) Bedrijven in Noord-Brabant*. Report for Brabant Development Corporation. Faculty of Geosciences, Utrecht University.

Keeble, D. and Nachum, L. 2002. Why do business service firms cluster? Small consultancies, clustering and decentralization in London and southern England. *Transactions of the Institute of British Geographers*, 27 (1), 67–90.

Koch A. and Stahlecker, T. 2006. Regional innovation systems and the foundation of knowledge intensive business services. A comparative study in Bremen, Munich and Stuttgart, Germany, *European Planning Studies*, 14 (2), 123–46.

Lee, C-K. and Saxenian, A. 2008. Coevolution and coordination: a systemic analysis of the Taiwanese information technology industry. *Journal of Economic Geography*, 8, 157–80.

Madhok A. and Liu, C. 2006. A coevolutionary theory of the multinational firm. *Journal of International Management*, 12, 1–21.

Muller, E. and Doloreux, D. 2009. What we should know about knowledge-intensive business services. *Technology in Society*, 31, 64–72.

Muller, E. and Zenker, A. 2001. Business services as actors of knowledge transformation: the role of KIBS in regional and national innovation systems. *Research Policy* 30, 1501–16.

OECD 2005. *Growth in Services. Fostering Employment, Productivity and Innovation*, Paris: OECD.

Oetzel, J. and Doh, J.P. 2009. MNEs and development: a review and reconceptualization. *Journal of World Business*, 44, 108–20.

Paterson, S.L. and Brock, D.M. 2002. The development of subsidiary-management research: review and theoretical analysis. *International Business Review* 11, 139–60.

Safarzynska, K. 2009. *Evolutionary Modelling of Transitions to Sustainable Development*, PhD-thesis, Amsterdam: VU University Amsterdam.

Schmid, S. and Schurig, A. 2003. The development of critical capabilities in foreign subsidiaries: disentangling the role of the subsidiary's business network. *International Business Review*, 12, 755–82.

Simmie, J. and Strambach, S. 2006. The contribution of KIBS to innovation in cities: an evolutionary and institutional perspective. *Journal of Knowledge Management*, 10 (5), 26–40.

Strambach, S. 2008. Knowledge-Intensive Business Services (KIBS) as drivers of multilevel knowledge dynamics. *International Journal of Services Technology and Management*, 10 (2/3/4), 152–74.

Taylor, P. 2004. *World City Network: A Global Urban Analysis*. London: Routledge.

Wang, J., Liu, X. and Li, X. 2009. A dual-role typology of multi-national subsidiaries. *International Business Review*, 18, 578–91.

Wood, P. 2009. Service competitiveness and urban innovation policies in the UK: the implications of the 'London Paradox'. *Regional Studies,* 43 (8), 1047–59.

Yeung, H.W.C. 2009. Regional development and the competitive dynamics of global networks, an East Asian perspective. *Regional Studies,* 43 (3), 325–51.

Chapter 9

# Regional Growth Dynamics: Intra-Firm Adjustment vs. Organizational Ecology

Harald Bathelt and Andrew M. Munro

## Introduction

The economic success of industries and regions depends on the capability of firms to develop structures of production and innovation that meet customers' demand and adjust to changes in the competitive and technology environment. As such, continued economic success requires different forms of corporate adjustments associated with 'industrial transition' – understood as changes in the technical, social and spatial divisions of labour (Massey 1984). In general, we can distinguish between different forms of transitions. On the one hand, there are constant incremental adjustments in production and efforts to make slight changes in products and innovation as a consequence of ongoing intra- and inter-firm learning processes. These become routine adjustments that do not require changing existing structures or question established technological trajectories. On the other hand, history shows that shifts in the competitive framework of an industry can also be of an unexpected, sudden or more drastic nature requiring substantial changes in corporate production and innovation philosophies (see, for example, Bathelt and Boggs 2003). Such adjustment pressures may be the result of a non-routine crisis such as the financial crisis in the late-2000s that shook the foundations of the global financial markets and severely affected production conditions in manufacturing. More radical adjustment or restructuring processes may also be necessary due to shifts in global competition or technological regimes. Although these shifts may have existed for a while, the need for drastic changes may not be obvious at first but may at some point become urgent and require fast adaptation and innovation. From the perspective of economic geography, a key question is how regions can support these transition, restructuring and adjustment processes to strengthen regional growth dynamics.

In this context, much of the literature focuses on regional clusters and innovation networks that have a specific industrial specialization (Maskell and Malmberg 1999, Cooke and Morgan 1998), while so-called 'normal regions' (Storper 1997) that are characterized by a diversified industry structure or no industrial focus at all have been neglected. Conceptual understandings in economic geography and regional economics are limited with respect to the conditions under which such regions may grow and how they adjust to new conditions in their competitive and technology environment. The related-variety approach that was developed in recent years addresses these questions by providing an explanation that emphasizes the role of cross-sectoral linkages in regional growth (Frenken et al. 2007). Accordingly, regions with a wide range of industries enable local firms to diversify their markets by tapping into the varied set of resources and competencies available locally. The approach suggests that firms can especially benefit from other local industries if their knowledge bases are overlapping and related to one another.

This chapter suggests a different approach that focuses on intra-firm adjustment and firm formation processes in response to changes in the corporate environment. In doing so, this chapter draws on findings from the organizational ecology literature (Hannan and Freeman 1993), combining it with findings from the literature on interactive learning and restructuring (Massey 1984, Lundvall 1988, Gertler 2004). In particular, we are interested in the role of firm formation, intra-firm restructuring and the relationship between both processes in stimulating technological change and growth at the regional level.

In our empirical analysis, we investigate the structure of production and innovation in the Kitchener and Guelph metropolitan areas, about 100 km west of Toronto, around which the initiative 'Canada's Technology Triangle' (CTT) was founded in the late-1980s. CTT was jointly established by the four cities of Cambridge, Guelph, Kitchener, and Waterloo to market the region's technological strengths and reduce inter-municipal competition. Although Guelph left this initiative later to market its strengths individually, the local economies are still linked to one another, drawing from a shared labor market and depending on similar economic conditions (Bathelt et al. 2010). Since the 1970s, this region experienced the impetus of firm formation processes from university research, particularly around the activities of the University of Waterloo. Numerous firms in the area of information technology (IT) were successfully launched, establishing a growing technology base in the region (Colapinto 2007, Bramwell et al. 2008). The region drew a lot of attention from policy makers because it was able to shift its economic focus from traditional industries to new IT-related businesses. Through media reports and academic studies, a regional success story of high-technology growth and university spin-off processes developed over time.

Since the 1970s, the regional economy achieved above-average performance levels, according to indicators such as job growth, unemployment or average household income. The unemployment rates, for instance, were among the lowest in any Canadian metropolitan area. In 2006, the unemployment rate in the Kitchener CMA and the Guelph CA was 5.6 and 5.1 per cent, respectively, compared to 6.4 per cent in Toronto and 6.1 per cent in Canada overall (*Statistics Canada* 2006). This supports the view that CTT benefited from university spin-offs and related transformation processes, not just giving rise to the IT industry but also refreshing the knowledge-base and reinvigorating firms in traditional manufacturing sectors.

On closer investigation, however, knowledge behind the success in regional growth and innovation appears limited. Clearly, the region cannot be viewed as a true regional industry cluster of closely interrelated firms of a particular value chain, and their supplier and service infrastructure (Bathelt et al. 2011). What we find instead is a diversified and segmented industry structure with businesses which are characterized by limited commonalities. The region hosts a substantial variety of large and small establishments, old and young firms, and businesses with diverse traditional and novel manufacturing and service backgrounds.

In the media hype around the supposedly 'post-industrial' future of the Waterloo region, it is often forgotten that the region also has a strong established manufacturing tradition (Rutherford and Holmes 2008). Both academic and policy analyses of innovation, however, tend to focus on high-technology growth and ignore the often informal innovation that happens in traditional industries. Upon closer analysis, we see that about 44,000 of 76,500 manufacturing employees (57 per cent) and about 1,050 of 2,150 establishments (48 per cent) in the region were within traditional manufacturing in the plastics and rubber, metal fabricating and processing, machinery, electrical equipment and automobile supplier industries in 2008 (Bathelt et al. 2012). In the Kitchener CMA, manufacturing had a share of 20.3 per cent of the total labor force in 2007, which was nearly twice as high as the Canadian average. As such, its contribution – despite its relative decline in the past decade – cannot be neglected in the economic success story of the region. The resilience of manufacturing in the region was tested severely during the global financial crisis in the late-2000s. Although there were significant job losses in 2008 and 2009, the region remained vibrant and recovered many jobs through 2010 (*Statistics Canada* 2009, 2010).

To explain CTT's economic growth dynamic, a study was undertaken to investigate the structure of and linkages in regional innovation and firm formation processes in both 'new economy' and 'old economy' sectors, as well as analyze potential relationships between both. This was based on the premise that regional growth dynamics and technological change are greatest if intra-firm adjustment processes in traditional manufacturing industries are linked with technology-oriented start-up processes in new sectors. Similar to the related-variety thesis, the hypothesis behind this is that a link between established industries and new technology-based firms may be particularly important to trigger regional innovation and growth. This relationship can be beneficial for both industry segments because it may trigger technological change in the former and provide early market access in the latter group of firms. According to this conceptualization, a study of innovation processes and production linkages in CTT was conducted in two stages: focusing (i) on IT-related university spin-offs and start-ups and (ii) on traditional manufacturing firms.

Accordingly, this chapter is structured as follows. In the second section, we discuss the conceptual framework revolving around ideas from organizational ecology, the role of spin-off firms and corporate adjustment processes. Extending common views in organizational ecology, we argue that start-ups in a regional context have the potential not only to trigger technological change themselves, but support and guide restructuring and modernization of established firms in other industries through learning networks. The following section presents the regional context of CTT, before we discuss the research approach and methodology applied. The sections of the empirical part analyze the nature of innovation linkages of both university-related IT spin-offs and start-ups and traditional manufacturing firms. The final section summarizes the main findings and draws some policy conclusions. It is suggested that our re-conceptualization may provide an alternative basis for regional policy approaches in the future, even though the investigation does not provide evidence of strong inter-sectoral linkages in our case study region.

## Organizational Ecology and Regional Technological Adjustment as a Selection Process

A conception which powerfully describes industrial transformation and emphasizes the role of young start-ups and spin-offs in organizational and technological change is that of organizational ecology (Hannan and Freeman 1977, 1993). Organizational ecology explains changes in organizational forms (firms) in an industry as an evolutionary process and claims that this change primarily results from selection processes among organizations, rather than from adjustments of existing organizations. In this approach, firms are viewed as not being easily structurally adaptable (Hannan and Freeman 1977). The adjustment to new market and technology conditions is seen as a process of selecting new products and technologies that meet the new conditions in the production environment best, thus providing higher returns than alternative choices. It is argued that power structures and routines within existing organizations provide obstacles for adjustments to new competitive structures because they require changes which are not easy to conduct – especially when these are not incremental in nature (Hannan and Freeman 1993). According to this approach, one reason for this is that intra-firm decisions are made on a consensual basis between different corporate interest groups, tending toward compromises which result in suboptimal structures. A second reason is seen in the slow pace of intra-firm adaptations because these might involve lengthy discussions and internal power struggles. In addition, it would be generally difficult to identify the most efficient adaptations because of existing uncertainties. All of these obstacles support structural inertia and make it difficult to adjust to new market and technology conditions in an effective way. In contrast, new firms are not affected by pre-existing structures that are bound to prior environmental conditions, but direct their new product and innovation structure to meet the present conditions.

As intra-firm adjustments and selection processes tend to be slow and sub-optimal, the organizational ecology literature argues that firms themselves become the object of market selection. The selection of firms is assumed to depend not only on their economic efficiency, but also on higher or lower reliability and accountability which result in a higher or lower degree of acceptance by potential buyers and users (higher or lower market legitimacy). In other words, this conception suggests that firms which produce reliable high-quality outputs and are easily accountable for their outputs will be more likely to gain a broad customer base and have market success. Hannan and Freeman (1993) argue that, even though selection processes prioritize firms with high reliability and accountability (i.e. well-established firms that are capable of constantly reproducing their own structure), new firms would be in a favorable position to survive in periods of organizational/technological change because they meet new requirements better and faster than established firms. Although new firms are characterized by relatively high exit rates, some have a greater robustness and will be successful. They will grow fast while established firms with outdated structures may disappear from the market due to structural inertia. As structural inertia increases with size and age, processes of adaptation in existing organizations become more difficult over time. Progress in major corporate restructuring processes would likely be slow, involve high switching costs and threaten reliability and market legitimacy, related to internal tensions and power struggles. In the end, it would be primarily new firms that drive new development trajectories. This idea has been utilized in the so-called density-dependence model to demonstrate how firm formation processes drive regional development, based on technological opportunities and market/ client legitimacy (Baum and Oliver 1992, Staber 1997). In that model, increasing market legitimacy of a small population of new firms specialized in a market niche may initially encourage further start-up processes and stimulate clustering and regional growth. At a later stage, however, increasing competition may lead to reduced start-up rates and slower regional growth, and a firm's legitimacy may eventually become key to its survival.

Although the organizational-ecology approach is particularly useful for pointing out the key roles of new start-ups in processes of technological change on the regional scale, it has been criticized for its neglect of (i) the role of agency, and (ii) the significance of permanent adjustment and learning processes in reacting to changes in the economic environment (for a summary, see Kieser and Woywode 1999). Many sectors are dominated by large firms which have existed and prospered over a long time period. These have been able to adapt new structures through ongoing interactive learning (Lundvall 1988, Gertler 2004), and to create their own regional environments (Storper and Walker 1989) by inducing regional suppliers to adjust their operations to service these large firms. Still, the importance of technology-based start-ups and university spin-offs in processes of technological change cannot be ignored (Vohora et al. 2004). They have the potential to become triggers of technological change as they develop their ideas for new products and services from basic and applied research, often not primarily driven by returns-to-investment considerations. The potential to develop new technologies is likely greatest if the universities or research facilities are specialized in particular science and technology fields, as in the case of the University of Waterloo. Aside from economic efficiency, the development of reliability and accountability is, however, a major obstacle for these spin-off firms as they need to find a home market and customer base to survive. They are typically founded by faculty members or graduate students with little experience in marketing, setting up production and establishing routines. This may, in turn, reduce their initial performance and regional impact in terms of jobs and limit their chances for survival.

In an attempt to combine the technological potential of new organizations with the effects of ongoing incremental adjustment and learning processes in existing organizations, we suggest an approach that opens up the deterministic logic of organizational-ecology models (Figure 9.1). Interestingly, the starting point of Hannan and Freeman (1977) was not much different from this model in that they assumed a combined process of corporate reorganization and firm formation. They were very critical though of the dominant model of corporate adaptation at that time, and thus focused on the role of new firms in organizational and technological change. In applying a spatial perspective, we argue in our model that technology-based spin-off processes have the potential to become a key driver of regional change. Similar to the ideas expressed in the density-dependence model (Baum and Oliver 1992), the spin-offs themselves can develop into a new basis for regional economic growth and trigger organizational and technological change in the first stage. This depends, of course, on their market legitimacy, and whether they develop reliable and accountable structures that produce economically successful products and services. As this seems to be a challenge for university spin-offs (Vohora et al. 2004), we can expect that the overall effects of technology-based start-ups are initially relatively small. This stage can take a long time as structures develop only incrementally, resulting in limited growth in employment and sales. If, however, the new firms become well established in their regional environment, generate production and research networks and turn into role models for others in adapting new technologies, they may have a strong overall impact in the region. This second stage is entered if regional interaction, input-output linkages and knowledge exchange stimulate adaptation processes within existing firms (Lundvall 1988). So, rather than being merely fitter than established firms, start-ups might induce learning processes in these firms. Since the established firms in the region have a strong record of efficient, reliable and accountable structures, their adaptation processes likely have a stronger regional impact in terms of additional incomes and jobs. The increased fitness of existing firms might, in turn, strengthen the legitimacy of new firms, provided that the latter establish a regional customer basis and learn to establish routines and institutions that strengthen reproductivity. In short, the combined potential of new firms and reorganized established firms can potentially stimulate broad learning and innovation processes at the regional level, eventually impacting other industries in the regional economy and stimulating further start-ups. Figure 9.1 illustrates the potential that may result from interdependencies between adaptation processes of established firms and technological triggers through new ones.

This is, however, not a deterministic process. We know that the successes of start-up firms and regional renewal processes depend on different factors, two of which deserve particular attention in our context: first, we know from empirical

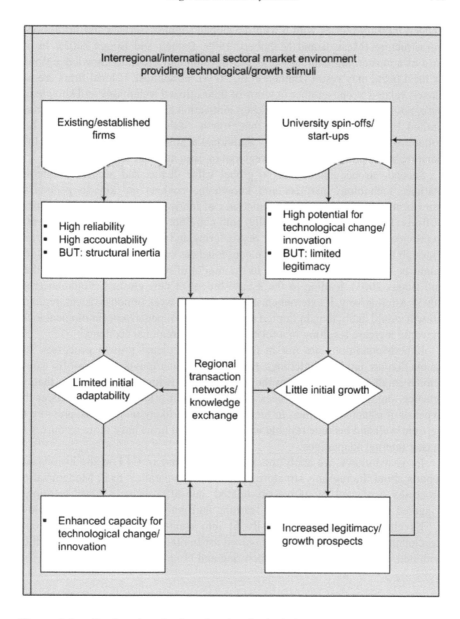

**Figure 9.1    Regional ecologies of technological change and growth**

studies that broad regional effects are more likely to occur if the regional economy draws from technological complementarities and overlapping knowledge bases which enable firms to establish regional networks and engage in knowledge exchange (Boschma 2005). A corresponding regional ensemble of firms can have the form of a regional thickening of a particular value chain or of a fully-fledged industry cluster with a well-developed supplier, service and institutional infrastructure (Maskell and Malmberg 1999, Bathelt and Boggs 2003). In the case of a cluster, regional networks can develop and dynamic knowledge flows, or local buzz, may result (Bathelt et al. 2004). In contrast, if local firms are not closely related to one another in terms of their utilized technology and knowledge base, possibilities for local networking, innovation and growth will likely remain limited. Instead, we might observe that growth – if there is growth at all – has few collective qualities but is caused by individual firm successes which rely on close bonding with partners outside the regional or even national economy.

Second, strong connections to global value chains and access to external markets, technology partners and knowledge pockets are key to generating growth impulses and sustaining processes of innovation in a regional ensemble of firms (Bathelt et al. 2004). Ideally, both conditions enable university spin-offs to provide ideas and incentives for restructuring and renewal in established firms. Through this combination, negative consequences of structural crises in a region might be overcome, as suggested in the model of regional re-bundling (Bathelt and Boggs 2003), leading to the establishment of new clusters (Feldman et al. 2005). Alternatively, if external cluster linkages are weak or non-existent, regional growth would likely remain limited and the danger of negative technological lock-in would increase resulting in a relatively unstable regional economy.

If both conditions are not in place, strong regional growth processes and innovation are unlikely outcomes. Initially promising university spin-offs would remain small or disappear from the market. What is important to note is that, in practice, ongoing selection processes likely do not prioritize small firms at the expense of established firms. In fact, spin-offs are likely under more pressure to perform well and become reliable while established firms have more resources to master internal adaptations.

In what follows, we apply this model to the case of CTT as the history and reports about the region's structure show stunning parallels to it. Modernization processes instigated by university-related spin-off processes have seemingly triggered regional modernization, learning and innovation. As described below, CTT's characteristics do not easily fit into common stereotypes of regional development as it is only a medium-sized urban region that is broadly diversified with both strong traditional manufacturing and IT-spin-offs firms.

**Regional Context of CTT**

The Kitchener and Guelph metropolitan areas were traditionally – and still are – characterized by a diversified manufacturing base. In the first half of the 20th century, the region had well-developed economic strengths in the rubber, textile, leather, furniture and food processing industries. Despite the differentiated industry structure, however, regional supplier linkages never seemed to be very strong (Bathelt et al. 2011). In the post-World War II period, manufacturing growth was driven by industries, such as fabricated metals, machinery, electrical products and the automobile supplier and transportation equipment sector (Rutherford and Holmes 2008).

Since the 1970s, a growing number of university spin-offs were started up in the region. This was related to the foundation of the University of Waterloo in 1959 as a university with an engineering focus and its intellectual property regime, allowing members of the university to own patents from university research. Industrial leaders of the region played an important role in the design of the university as they shaped the university's co-operative education program, its openness toward private sector collaboration and focus on applied research (Colapinto 2007, Bramwell and Wolfe 2008). The co-op program and a constant flow of highly qualified graduates were significant in supporting economic growth in the region. Aside from start-ups around the University of Waterloo, the region also attracted a number of multinational IT firms that established branches or acquired existing technology firms, thus forming a growing technology sector.

New technology-based firm formation played an important role in supporting the transformation of the region's manufacturing base, while remaining strongly diversified. Traditional manufacturing sectors such as textile mills, clothing and leather manufacturing lost between 50 and 60 per cent of their employees in the Kitchener CMA between 2001 and 2006; and chemical and electrical equipment manufacturing lost 20 per cent of their employees. This structural change was compensated by a 20 per cent increase of employment in plastics/rubber product and computer/electronic product manufacturing. Most knowledge-based producer-related services also experienced substantial job growth (Bathelt et al. 2011). As a result, the regional unemployment rates were among the lowest in Canadian metropolitan areas.

This situation changed drastically with the global financial crisis of 2008–09, which was unique in both its speed and severity. The crisis had a significant impact on the region's economy due to its dependence on traditional manufacturing. The Kitchener and Guelph metropolitan areas were strongly hit as indicated by shifts in unemployment figures. In March 2009, the unemployment rate in the Kitchener CMA increased to 9.6 per cent, exceeding the respective figures of the Toronto CMA (8.8 per cent) and Canada (7.6 per cent). The region lost over 10,000 manufacturing jobs since early 2007, representing more than a 17 per cent loss in this labor market segment (*Statistics Canada* 2009). While the economy experienced a strong downturn, the question arose whether established manufacturing sectors were strong and innovative enough to overcome this crisis. As it turned out (and as far as we can say at this point), the region recovered quickly and between August 2009 and 2010 about 8,500 new manufacturing jobs were created in the Kitchener MSA. As a consequence, the unemployment rate decreased to 7.0 per cent (*Statistics Canada* 2010), below that of the Toronto MSA (9.1 per cent) and Canada overall (8.0 per cent).

As already mentioned above, our understanding of regional growth and renewal processes still appear limited. There is no simple explanation for the long-lasting regional success and the fast recovery after the global financial crisis. In fact, a number of questions remain unanswered that require our attention: how, for instance, were traditional manufacturing industries able to regain their strength after the crisis? Did they benefit from ongoing linkages with the region's growing technology base and the technology-based start-ups that had been formed before? And, did university start-ups and spin-offs, in turn, benefit from a regional market that was open to modernization? Finally, did regional production and innovation networks develop or help overcome the crisis effects? To answer these and related questions, we conducted a qualitative study focusing on innovation practices and production networks in two core segments of the regional economy.

## Methodology

To investigate our hypothesis regarding the triggers and backgrounds of regional growth and innovation, we designed a qualitative study that looked into the structure of innovation and firm formation processes in two phases. In the first phase in 2007/08, we conducted semi-structured interviews with 18 IT start-up/ spin-off firms from the University of Waterloo. The total number of IT start-ups/ spin-offs identified in the region was 42 (Bathelt et al. 2010). Of these, 32 firms were contacted resulting in 14 rejections (44 per cent) and 18 interviews. In the second phase in 2008/09, another 40 firms in traditional manufacturing industries were interviewed. Of a total of 642 traditional manufacturers identified in selected industries (Bathelt et al. 2012), 310 were contacted and asked to participate in our study. Of these, 270 rejected and only 40 agreed to a personal interview. The unusually high rejection rate of 87 per cent is indicative of the difficulties we encountered in engaging firms in this research in the midst of a severe economic-financial crisis. Additionally, we conducted 8 interviews with university technology transfer officers, economic developers of the cities and leading representatives of business organizations. The latter interviews were conducted for the purpose of triangulation and getting an overview of the overall start-up policies and innovation dynamics in the region. The interviews took on average about one hour and were recorded on tape (Bathelt et al. 2011).

Our interviews focused on regional firm formation and innovation processes in both new and old economy sectors. We investigated the way in which firms established regional linkages or networks in innovation, to which degree they developed and depended upon global linkages, or pipelines, and whether this dynamic produced spillovers to other regional industries, due to practices such as inter-sectoral networking, technology transfer and job hopping. Through this, we aimed to explore the processes of adjusting to or adapting to changing competitive and technology conditions at the corporate level and how this was linked to broader regional renewal and growth processes. In particular, our goal was to investigate whether our organizational-ecology model that combines firm formation with intra-firm adjustment processes helps to understand the regional growth dynamics in CTT.

During the interviews, we primarily talked to the founders, executives or chief operational managers of the firms. Our questions focused on three main areas of interest: first, we asked with what goals and incentives and under which conditions the firms were started up in the region. The second set of questions was concerned with material linkages and knowledge flows related to innovation, both within the region or with partners in other regions and countries. Third, we were interested in finding out whether local institutional support and economic policies provided incentives to develop local linkages and how important cross-industry linkages were in innovation. Results of this research have been published in a number of qualitative studies (Bathelt et al. 2010, 2011, 2012). In the remainder of this chapter, we present a synthesis of this empirical research conducted with a different conceptual focus, and without presenting detailed individual-firm case studies.

## University-Related IT Start-Ups in CTT

CTT is frequently portrayed as a dynamic technology region that draws from university start-up/spin-off processes, knowledge transfers and corresponding regional networks. Our research, however, portrays a different picture of the development in this region and offers interesting insights into the underlying social dynamics of innovation processes. In contrast to what we expected, we did not find proof of strong value-chain based or cross-sectoral networks and knowledge flows in innovation. Overall, we also found little evidence of dynamic interactions of technology-based start-up firms with existing regional industries. The challenge of achieving market legitimacy of the new firms was usually solved individually through linkages with customers and partners in other regions and countries.

We started off with the assumption that IT firms would be the most likely of the university start-ups that would demonstrate evidence of important regional linkages in innovation, both within and across value chains. However, the empirical results derived from our interviews showed that, first, there were fewer such start-ups and spin-offs than expected and, second, most of these firms operated in specific cross-regional networks along market and technology linkages that adhere to their particular technological expertise. Local linkages with customers and suppliers and the existence of regional innovation networks were quite limited in their extent or absent altogether (for a different case study, see Rees 2004). This is clearly indicated in Tables 9.1 and 9.2. Only about one-third of the IT firms interviewed had significant local supplier linkages ($\geq$ 10 per cent of overall supplies), and just one firm purchased its supplies primarily locally (Table 9.1). In terms of sales linkages, only one firm showed a significant local orientation ($\geq$ 10 per cent of overall sales), while most had predominantly international customer linkages (Table 9.2). Almost 90 per cent of the IT start-ups/spin-offs sold most of their outputs in the United States or overseas. While these patterns do not focus on innovation linkages *per se*, the patterns were still reflective of the nature of knowledge flows in innovation which, according to our research, primarily involved non-local producer-user linkages (Bathelt et al. 2010, 2011). Although some of the start-ups and spin-offs had achieved a high degree of market legitimacy in their business segments, they were not well linked with other firms in the region.

**Table 9.1    Local supplier linkages of firms in Canada's Technology Triangle, 2007–2009**

| Firm type | Firms with significant/high local supplies | | | |
|---|---|---|---|---|
| | Significant (≥ 10 %) | | High (≥ 50 %) | |
| | Number | Share | Number | Share |
| IT start-up/spin-off firms | 6 of 17 | 35 % | 1 of 17 | 6 % |
| Traditional manufacturing firms | 14 of 37 | 38 % | 11 of 37 | 30 % |

*Source*: Survey Results

**Table 9.2    Local and international sales of firms in Canada's Technology Triangle, 2007–2009**

| Firm type | Firms with significant/high local sales | | | | Firms with significant/high international sales | | | |
|---|---|---|---|---|---|---|---|---|
| | Significant (≥ 10 %) | | High (≥ 50 %) | | Significant (≥ 10 %) | | High (≥ 50 %) | |
| | Number | Share | Number | Share | Number | Share | Number | Share |
| IT Start-up/ Spin-off Firms | 1 of 15 | 7 % | 1 of 15 | 7 % | 14 of 15 | 93 % | 13 of 15 | 87 % |
| Traditional Manufacturing Firms | 18 of 37 | 49 % | 8 of 37 | 22 % | 23 of 37 | 62 % | 14 of 37 | 38 % |

*Source*: Survey Results

Regional industry organizations were also of limited importance in stimulating innovation, primarily playing a role in deepening social networks and generic skill sets (see also Bramwell et al. 2008). Firms frequently turned to specialized Internet-based user groups as their initial problem-solving tool and rarely found opportunities to collaborate with other firms in the region, typically citing that nobody else was working on the same type of products and problems. It is unlikely that these firms could spur the development of specialized regional innovation networks.

Although the IT sector may be somewhat specific in terms of its ability to create international networks, it does not possess fundamentally different linkage patterns compared to other new technologies. In particular, we expected university start-up firms to display a stronger regional orientation in their early stages. This was, however, not the case. We found three reasons that help explain this: first, it seemed that firms in the area of specialized software solutions were able to establish a broader extra-regional customer base more quickly and easily than firms in other sectors. Second, the regional firms were extremely diversified, limiting opportunities for local network creation in a mid-sized region. Third, acquisitions of firms by larger entities that took place in the region served to provide access to wider extra-regional corporate networks, and thus boosted market legitimacy for the respective units.

We encountered only a few IT start-ups and spin-offs that reported linkages in innovation with firms in other sectors, and none of these firms cited a single particular industry outside their value chain as important for innovation. One producer of electronic control boards identified some of the manufacturers interviewed, but did not name any of those firms as an important customer or partner in innovation. In a second case, a firm sold its CNC software to a local manufacturer, but distributed the software through an out-of-region machining hardware supplier. The regional connection was merely accidental and the interviewee only knew about it through a social relationship. When asked about the value of cross-sectoral relationships, the firms did not indicate that these were important. The answers received were usually quite generic, with firms most commonly responding that a diversified economic base would provide a diversified labor market. However, nothing specific was mentioned about the value to their operations, suggesting that such linkages were not very common. The absence of noticeable cross-sectoral linkages in innovation clearly indicated that implicit claims about regional spillovers from regional IT start-ups to traditional manufacturing firms may be over-stated.

With respect to the conditions for regional innovation, we found that most firms were stand-alone units in the regional economy with strong international linkages, particularly to the United States. With the exception of those firms that had a hardware-related component to their product offering, they had little ongoing research activities with R&D laboratories and the regional universities. Our study indicated that limited specific knowledge was transferred to regional industries through entrepreneurial faculty members and graduates. Generally, the firms interviewed were able to create legitimacy in their industrial sectors and establish close customer relationships, but these developed largely external to the region. They offered little in the way of adaptation stimuli for established firms in other regional industries and thereby did not induce broader region-specific innovation ad transformation processes.

## Traditional Manufacturing Firms in CTT

In the second phase of our research, we interviewed firms in traditional manufacturing industries in the areas of plastic and rubber products, fabricated metal products, machinery, electrical equipment and transportation equipment. We were especially interested in the structure of innovation processes and the potential linkages between traditional manufacturers and university-related IT start-ups in the sense of the model in Figure 9.1, as these two segments – compared with others – were most likely to collaborate due to a potential overlap in knowledge bases.

Our results indicated that the firms interviewed differed widely in their role in the design process of the products they fabricated or manufactured. Some firms primarily performed generic treatments, such as heat-treating or painting, to the products. Similarly, contract fabrication shops, which conducted limited runs of machining and/or CNC manufacturing from a client's design, often had little involvement in the development of the products they produced. Nonetheless, some of the firms developed internal capacities to take the designs and provide input to their customers from a manufacturing stand-point. Over time, some firms became increasingly involved in early-stage design processes of the end products. We also encountered firms in the upper tiers of the manufacturing value chain that both manufactured and designed products. Overall, we were surprised that most of the manufacturing firms interviewed were quite innovative in recent years: one third had developed new products and another third introduced new processes in the two years prior to our interviews.

Firms in traditional manufacturing differed substantially in terms of supplier-customer relationships and the kinds of knowledge-transfer-based innovative activities they engaged in, depending on where they were positioned in the value chain. Altogether, we identified two distinct cohorts of firms in our sample that were active in innovation. One cohort primarily engaged in custom manufacturing, converting raw materials into finished components using other firms' designs. They worked from a blue-print and gave feedback primarily on the manufacturability of the parts, rather than their end function. For these firms, most of their innovation came from cost-cutting measures and improving workflow and material management to reduce production delays. Other sources of innovation were capital investments in new equipment in order to increase capabilities and enhance automation to save on labor costs. The second cohort generally utilized their own designs to fabricate products, even if their outputs fed into other firms' product value chains. Some of these firms were very innovative with long-term research strategies in the development of new products. This was usually related to close interaction with customers, which were often not nearby.

With an opportunity to observe different cohorts of firms, we expected to see that the underlying innovation networks would be based on regional collaboration. However, it turned out that the firms interviewed usually did not identify either regional suppliers or customers as important to their innovation processes. Nonetheless, local/regional linkages through purchases of supplies and sales of products were higher than in the segment of IT start-ups/spin-offs: almost 40 per cent of the firms purchased at least some raw materials ($\geq$ 10 per cent of overall supplies) from within the wider region (14 of 37, see Table 9.1), but their decision to do so was primarily based on logistical concerns and price factors, and not on innovation inputs and learning effects. Compared to the IT firms, a higher share of traditional manufacturers had primarily a regional supply orientation ($\geq$ 50 per cent of overall supplies). In terms of sales linkages, half of the traditional manufacturers had at least some significant local sales and about one fifth were primarily locally oriented (Table 9.2). Given that we interviewed numerous tier-2 and tier-3 manufacturers, fabricators and metal treaters, which would typically try to access local markets and provide customized jobs, this orientation toward local sales appeared lower than expected. In fact, a substantially higher share of firms was mostly reliant on customers from the United States or Europe: almost 40 per cent of the firms had primarily international sales. Again, although these numbers do not specifically relate to innovation, they reflect the linkage patterns observed in innovation, which, according to our interviews, often did not have a strong regional component (Bathelt et al. 2012).

Firms overwhelmingly indicated that ideas for new products were developed internally, sometimes through corporate ties connecting with facilities in different countries. Most firms did not have important regional suppliers or customers they collaborated with in innovation; thus, it was not surprising that these firms relied on internal problem-solving. The firms also did not use consultants or have close relationships with research laboratories. When they ran into problems in their production processes, the way they responded varied depending on the type of firm: for the metal fabricators and treaters, the driving forces for innovation were customer demand to lower costs and shorten turn-around times. For OEMs, the expectation of customers was related to improved product performance or design and feature enhancements. Regardless of the firm type, the customers were seen as the main push behind innovation, and often the key source for design input. Yet, these influences were largely outside the region. Although sales in traditional manufacturing sectors were not as internationally oriented as in the IT segment, such linkages were critical to the success of these firms.

As in the case of IT start-ups/spin-offs, we did not find strong evidence for cross-sectoral linkages in innovation and learning. Most commonly, firms indicated that they benefited from a joint labor pool in the region and the ability to draw skilled employees from firms in other industries using similar processes. But the flow of employees between firms was not seen as a significant input for innovation. A couple of firms indicated that they benefited from having a diverse pool of close-by suppliers, but suggested that the benefits came from the ease of having access to them, rather than from direct learning processes. Only one firm made explicit reference to business or management benefits, suggesting that – since so many firms in the region exported around the world – discussions with firms in unrelated sectors on how to solve export challenges and share information and strategies on business operations were useful. Most answers appeared vague, however, indicating that cross-sectoral linkages in innovation were not common.

Generally, the results of the second research phase on traditional manufacturing in CTT paralleled the observations about regional innovation processes from the first phase. Again, we did not see substantial patterns of local supplier-customer linkages that were important to innovation, learning and adaptation processes. Firms relied on international linkages, often with partners in the United States, and used regional suppliers primarily for generic business services, labor and raw materials. Most firms indicated that they had little employee turn-over, and many participated in the regional high school and community college apprenticeship programs, but had no distinct regional innovation networks otherwise. It turned out that traditional manufacturing firms were more willing to learn, innovate and adjust their structure to changing market needs than what a conventional organizational-ecology model would suggest. However, this willingness was not related to regional innovation networks and close linkages to technology-based IT start-ups as suggested in Figure 1; it was primarily a consequence of systematic learning and adjustment capabilities in intra-firm networks and through international market ties.

## Conclusions

The research presented in the chapter aims to understand the regional growth dynamics of CTT, a region 100 km west of Toronto, by analyzing its internal and external production and innovation linkages. Our arguments draw from a modified organizational-ecology conception that combines intra-firm adjustment with firm formation processes at the regional level. The basic idea of this approach suggests that technological change and regional growth, such as that witnessed in CTT, are most successful if technological triggers of start-ups and learning and adaptation processes of established firms are linked with one another and support each other. The model in Figure 9.1 suggests that the formation of regional networks in production and innovation has advantages for existing and new firms and reduces risks of market failures related to both limited legitimacy and adaptability. The transformation that has been observed in the region of Cambridge, Guelph, Kitchener and Waterloo from an economy based on traditional manufacturing to one with a substantial proportion of IT-related businesses is often attributed to knowledge transfers and growth triggers based on university spin-off processes and related innovation and learning networks – a process that would support the combined organizational-ecology and adaptation model. Accordingly, we expected to find evidence of close regional networks and systematic linkages between traditional and new technology-based industries.

Surprisingly, however, we did not find much evidence that such processes were at the core of regional growth dynamics. Our research showed that local firms were typically not closely related to one another in their technological and knowledge bases. This provided limitations for local networking, learning and knowledge flows between firms. In contrast, firms tended to engage in international linkages to provide the necessary growth impulses, both within corporate networks and through inter-firm linkages. Successful regional restructuring and modernization was primarily a result of individual-firm competencies, shared generic knowledge assets and a strong sense of community in marketing the region's attributes, rather than the effect of collective and cross-sectoral endeavors in innovation or networking. We found less than a handful of examples of some sort of relationships between traditional manufacturers in the region and IT start-ups/spin-offs, and where they existed the firms did not indicate that they were relevant for their learning and innovation processes. Of course, this should not be taken to suggest that there are no cases of cross-industry linkages, nor do we mean to suggest that such relationships have not been important for some firms outside of our sample; but our results show that cross-industry linkages between value chains are uncommon and that their influence on regional innovation processes appears minimal.

At the same time, our study showed that traditional manufacturing firms were not less resilient towards economic crises than new, creative or high-technology industries. Firms in the former industries were able to flexibly adjust to new market situations by the means of incremental improvements and adjustments, the acquisition of new resources and related diversification and renewal processes. There is little evidence suggesting that network qualities or cross-sectoral linkages in innovation – especially between established and new industries – were key to the success of these firms. It appears that regional growth dynamics have primarily relied on the wider internal corporate networks and firm-specific competencies, as well as intensive linkages with foreign markets – all of which in the context of a diversified regional economy and dynamic labor market.

These findings, of course, leave us with a puzzle regarding the combined organizational-ecology and intra-firm adaptation model: do these findings lead to the conclusion that our modified conception is unrealistic or irrelevant in our case study? We do not believe so, and suggest, instead, that the model can be quite useful from a regional policy perspective. This is because it focuses on a combined strategy to support both the strength of established and new industries, and to modernize the regional economy to reinvigorate growth. This strategy can be applied to normal regions not characterized by industrial specialization or to fully-fledged industry clusters, as long as they host segments of older established firms and young start-ups/spin-offs with related knowledge bases. Even in CTT, this model might be an interesting starting point for new policy initiatives as the current fragmented economy may experience strong growth in some periods (based on existing diversification advantages), yet it may under-perform due to a lack of economic cohesion and little collective synergies in innovation in other periods in the future.

Our model and empirical case study indicate that regional growth dynamics cannot be viewed independently from trans-regional or global market and technology linkages (Bathelt et al. 2004). Nonetheless, this suggests that regional policies may help strengthen learning and innovation capabilities. First, policy initiatives can aim to encourage technology-based start-up and spin-off processes to trigger regional innovation and support new firms to find initial market access. Second, regional policy initiatives can address corporate learning and adjustment capabilities in traditional industry segments and provide incentives for technological adaptation and innovation. Thirdly, programs can be developed that support the establishment of production, learning and innovation networks between technology-based start-ups and existing firms in more traditional manufacturing segments to provide early-stage market access for the former and incentives for modernization and innovation for the latter.

## Acknowledgements

We wish to thank Dieter Kogler, Ben Spigel and the book editors for their support and two anonymous reviewers for insightful comments.

## References

Bathelt, H. and Boggs, J. S. 2003. Towards a reconceptualization of regional development paths: Is Leipzig's media cluster a continuation of or a rupture with the past? *Economic Geography*, 79, 265–93.

Bathelt, H., Kogler, D. F. and Munro, A. K. 2010. A knowledge-based typology of university spin-offs in the context of regional economic development. *Technovation*, 30, 519–32.

Bathelt, H., Kogler, D., F. and Munro, A. K. 2011. Social foundations of regional innovation and the role of university spin-offs. *Industry and Innovation,* 18, 461–86.

Bathelt, H., Malmberg, A. and Maskell, P. 2004. Clusters and knowledge: Local buzz, global pipelines and the process of knowledge creation. *Progress in Human Geography*, 28, 31–56.

Bathelt, H., Munro, A. K. and Spigel, B. 2012. Challenges of transformation: Innovation, re-bundling and traditional manufacturing in Canada's Technology Triangle. *Regional Studies*, 46, forthcoming.

Baum, J. A. and Oliver, C. 1992. Institutional embeddedness and the dynamics of organizational populations. *American Sociological Review*, 57, 540–59.

Boschma, R. A. 2005. Proximity and innovation: A critical assessment. *Regional Studies*, 39, 61–74.

Bramwell, A., Nelles, J. and Wolfe, D. A. 2008. Knowledge, innovation and institutions: Global and local dimensions of the ICT cluster in Waterloo, Canada. *Regional Studies,* 42, 101–16.

Bramwell, A., Wolfe, D. A. 2008. Universities and regional economic development: The entrepreneurial University of Waterloo. *Research Policy,* 37 (8), 1175–87.

Colapinto, C. 2007. A way to foster innovation: A venture capital district from Silicon Valley and Route 128 to Waterloo region. *International Review of Economics,* 54, 319–43.

Cooke, P. and Morgan, K. 1998. *The Associational Economy.* Oxford: Oxford University Press.

Feldman, M., Francis, J. and Bercovitz, J. 2005. Creating a cluster while building a firm: Entrepreneurs and the formation of industrial clusters. *Regional Studies,* 39, 129–41.

Frenken, K., Van Oort, F. and Verburg, T. 2007. Related variety, unrelated variety and regional economic growth. *Regional Studies,* 41, 685–97.

Gertler, M. S. 2004. *Manufacturing Culture: The Institutional Geography of Industrial Practice.* Oxford, New York: Oxford University Press.

Hannan, M. T. and Freeman, J. 1977. The population ecology of organizations. *American Journal of Sociology,* 82, 929—64.

Hannan, M. T. and Freeman, J. 1993. *Organizational Ecology.* Cambridge (MA), London: Harvard University Press.

Kieser, A. and Woywode, M. 1999. Evolutionstheoretische Ansätze (Evolutionary approaches), in *Organisationstheorien (Organization Theories),* edited by A. Kieser. Stuttgart: Kohlhammer, 253–85.

Lundvall, B.-Å. 1988. Innovation as an interactive process: From producer-user interaction to the national system of innovation, in *Technical Change and Economic Theory,* edited by G. Dosi et al. London, New York: Pinter, 349–69

Rees, K. 2004. Collaboration, innovation and regional networks: Evidence from the medical biotechnology industry of Greater Vancouver, in *Proximity, Distance and Diversity: Issues on Economic Interaction and Local Development,* edited by A. Lagendijk and P. Oinas. Aldershot: Ashgate, 191–215.

Maskell, P. and Malmberg, A. 1999. The competitiveness of firms and regions: 'Ubiquitification' and the importance of localized learning. *European Urban and Regional Studies,* 6, 9–25.

Massey, D. 1984. *Spatial Divisions of Labor: Social Structures and the Geography of Production.* London: Macmillan.

Rutherford, T. and Holmes, J. 2008. Engineering networks: University-industry networks in Southern Ontario automotive industry clusters. *Cambridge Journal of Regions, Economy and Society,* 1, 247–64.

Staber, U. 1997: An ecological perspective on entrepreneurship in industrial districts. *Entrepreneurship and Regional Development,* 9, 45–64.

*Statistics Canada* 2006. Census of Canada 2006: Profile Series, Cumulative Files. Catalogue No. 94-581-XCB2006001 and 94-581-XCB2006004. Ottawa: Statistics Canada.

*Statistics Canada* 2009. CANSIM Tables 2820090 and 2055599. Labour Force Survey (LFS) Estimates. Ottawa: Statistics Canada.

*Statistics Canada* 2010. CANSIM Tables 2067453, 2067458 and 2067467. Labour Force Survey (LFS) Estimates. Ottawa: Statistics Canada.

Storper, M. 1997. *The Regional World: Territorial Development in a Global Economy*. New York: Guilford Press.

Storper, M. and Walker, R. 1989. *The Capitalist Imperative: Territory, Technology, and Industrial Growth*. New York, Oxford: Basil Blackwell.

Vohora, A., Wright, M. and Lockett A. 2004. Critical junctures in the development of university high-tech spinout companies. *Research Policy*, 33, 147–75.

Chapter 10

# Transnational Entrepreneurs and the Global Shift of Production: The Example of Diamond Manufacturing

Sebastian Henn

## Introduction

In the past decades significant improvements of transportation and communication technologies have given rise to the change of global migration systems and the corresponding development of a specific segment of small immigrant businesses which increasingly has aroused the interest of the scientific community (Sternberg and Müller 2010, Drori et al. 2009, Portes et al. 2002). What is meant are businesses established by self-employed migrants who keep in close touch to their peers, and, by falling back on their transnational social networks are able to gain access to new resources which endow them with significant competitive advantages in the global marketplace.

This type of cross-border business has mainly been discussed in the context of corporate and sociological studies (Drori et al. 2009) and recently has attracted also the interest of Economic Geographers. So far, the debate in the latter field has been dominated by two perspectives: On the one hand, different authors have analyzed the corporate internationalization strategies of transnational entrepreneurs. A clear focus has been put on the expansion of Asian firms in Southeast Asia as illustrated by the seminal works by Yeung (2007) and others (e.g. Hsing 1996). On the other hand, scholars have sought to understand how transnational entrepreneurs contribute to the transfer of knowledge from industrialized to developing or industrializing countries and thus allow for an upgrading of the respective peripheral economies. In this context, it was shown that skilled transnational entrepreneurs with working experience in Western leading-edge technology regions who eventually return to their home countries not only establish new global production networks but also contribute to the formation of new clusters in their home countries (Saxenian 2006, Sternberg and Müller 2010). Typically, these new centers have not been found to compete with existing clusters but rather to fulfill complementary functions (Saxenian 2006) which is not surprising when considering the fact that research has been confined to high-technology industries in many cases requiring capabilities developing and industrializing countries still lack (Saxenian 2006).

In fact, the one-sided consideration of knowledge-intensive sectors which heavily depend on leading-edge technologies must be considered as a major shortcoming of the present studies on transnational entrepreneurship. The engagement of transnational entrepreneurs in large-scale manufacturing with price competition being the most important competitive factor, for example, could result in different processes like relocations from industrialized to emerging economies with lower labor costs rather than in the emergence of complementary industrial structures. To contribute to our understanding of transnational entrepreneurship and to understand whether such processes actually occur, this study analyses how transnational entrepreneurs contribute to the change of spatial patterns in the manufacturing sector. Doing so, it will focus on the diamond cutting industry as a very labor-intensive industry which in large parts nowadays is still characterized by traditional manual working techniques.

The chapter has been structured as follows: The next section deals with transnational entrepreneurs from a conceptual point of view. It will be shown that the growing importance of these agents can be attributed to their ability to link different locations thereby optimizing their corporate production processes. The subsequent section aims at transnational entrepreneurs as elements of industrial transition: In particular, transnational entrepreneurs will be understood as agents being able to contribute to both the emergence of global production networks and new regional clusters. In order to study how transnational entrepreneurs affect clusters in the manufacturing sector, a qualitative framework has been applied which will be outlined in the next part. The following sections present the empirical results. At first, some basic features of the diamond sector will discussed. After that, the social characteristics of the Palanpuris, a relevant social group of actors in the diamond sector, will be introduced. Furthermore, it will be shown how the Palanpuris entered the market of diamonds resp. how they contributed to a global transfer of knowledge and developed transnational networks allowing for the evolution of new cluster structures which more or less outcompeted the once most important diamond center Antwerp. Finally, the last section draws conclusions and sheds light on further research issues.

## Transnational Entrepreneurs – A Conceptual Framework

For about ten years now, the concept of transnational entrepreneurship has developed as a new interdisciplinary research field at the interface between sociology, business economics and geography (see for example Drori et al. 2009, Yeung 2007, Portes et al. 2002). Historically, it can be traced back to the debate on transnationalism (Portes et al. 2002) which itself accrues from migration theory and accommodates the recent changes of global migration systems and their social and spatial implications.

A central aspect of the literature on transnationalism has been the fact that today's global migration processes cannot be simply considered as unique and unidirectional movements but rather must be understood as taking place in an oscillatory pattern between different locations. This development which has clearly been enforced by modern means of communication and transportation has gone along with the detachment of distinct social groups of actors from their ancestral locations; nowadays they simultaneously engage in two or more different environments and thereby contribute to the formation of so-called transnational social spaces. Usually, these agents are termed transnational migrants or shortly transmigrants with the attribute 'transnational' pointing at both their concomitant presence in different parts of the world and their transnational networks which connect them with relatives and other actors on a global scale (Henn 2010, Sternberg and Müller 2010).

Typically, transmigrants have been found to exhibit a clear tendency of being self-employed in international trade (Light 2010, Portes et al. 2002). With reference to their commercial activities, these 'self-employed immigrants whose business activities require frequent travel abroad and who depend for the success of their firms on their contacts and associates in another country, primarily their country of origin' (Portes et al. 2002: 287) are referred to as transnational entrepreneurs (Light 2010). Even though this definition captures a broad spectrum of different types of entrepreneurs (for an early classification of their companies see, for example, Landolt et al. 1999) all of them have at least one thing in common: Transnational entrepreneurs are in the advantageous position to link formerly unconnected networks of market agents which exist at two or more (distant) locations. In other words, acting in structural holes (Burt 1992), transnational entrepreneurs find themselves 'in a unique position to exploit opportunities either unobserved, or unavailable, to other entrepreneurs located in a single geographical location' (Drori et al. 2009: 1002). Of course, benefiting from this network position clearly presupposes certain characteristics of the involved actors: First, the entrepreneurs must have a certain foresight of venturing abroad and expanding their businesses beyond a 'comfortable home market share' (Yeung 2007: 579). Second, as the actors have to learn to cope with 'unexpected contingencies' (Yeung 2007: 577) and to understand the realities in their host countries informal information and support in the host country are of critical importance to them. This aspect points at the high relevance of their embeddedness in a local actor network which functions both as institutional basis for entrepreneurship and as necessary strategic infrastructure determining the success of transnational entrepreneurs (Yeung 2007). In fact, transnational entrepreneurs typically belong to closely knit groups of actors characterized by a common institutional background (e.g. religion) and trust networks (Coleman 1990) and thus on two factors which facilitate the exchange of information and artifacts (Hsing 1996). Besides ethnic groups, clans or joint families play an important role in this context (Bagwell 2008). Especially the latter have been found to provide entrepreneurs with distinct advantages in their internationalization endeavors as familial relationships are 'long-lasting and do not appear to need frequent contact in order to be activated' (Mustafa and Chen 2010: 104). In addition, they provide the entrepreneurs with emotional support but also allow for the access to specific resources like pooled savings, family labor or access to trusted networks and thus provide a strong basis for support not only during the emergent stage of a business but also during the development of the firm. In this manner, family links, for example, have been used for importing or exporting goods resp. financial and/or human capital (Bagwell 2008, Yeung 2007).

## Transnational Entrepreneurs as Elements of Industrial Transition

Recent studies in Economic Geography which refer to transnational entrepreneurship so far have mainly been confined to the 'New Argonauts' (Saxenian 2011, 2006, Sternberg and Müller 2010). This term has been applied to high-qualified migrants who for a considerable time have stayed abroad thereby acquiring new knowledge and contacts but someday realize that they can take advantage of their own networks to their countries for providing the (former) host country with lacking skills. Many migrants thus returned home and established new enterprises there. Even though the initial investments might have been driven by low labor costs, these companies in first instance distinguish themselves from other low-cost locations by collaborating with partners in their former host-country (Saxenian 2011). As a consequence of the entrepreneurs' activities, formerly distant regions of the world become linked to leading markets and technologies and thus to globally important sources of information, skills and tacit knowledge. Since the agents in question typically return to larger cities with developed research infrastructure and a diversified industrial basis, transnational entrepreneurs at the same time 'provide a mechanism for seeding entirely new centers of low-cost (at least initially) supply in less developed regions' (Saxenian 2002: 186) characterized by specialized skills and expertise.

Even though transnational entrepreneurs have been found to be important catalysts for cluster formation only little attention has been paid to the question how these newly emerging clusters interact with existing economic structures in the migrants' (temporary) host countries. So far, it has been suggested that the new clusters typically do not make up a strong competition to the existing ones. Saxenian (2011: 6) in her study on transnational entrepreneurs in the Silicon Valley, for example, states that they 'rarely compete head-on with established US producers; instead, they build on the skills and the technical and economic resources of their home countries'. Using the case of Thai Argonauts, she convincingly shows that their firms developed complementary specializations thereby forming a new global production network (GPN) between the US and Thailand. In case of entrepreneurs from China, Saxenian (2002) comes to an analogous conclusion when stressing that the majority of the Argonauts' mainland start-ups tend to locate their headquarters in the US while their operations are primarily carried out in China. Similarly, Sternberg and Müller (2010) prove that Chinese return migrants make use of advantages of different locations by breaking down their value-added chains to different locations, thereby setting up global production networks: The production of knowledge takes place in the US while the commercialization has been localized in China.

Overall, the contributions clearly underline the ability of transnational entrepreneurs to develop new GPNs marked by the integration of production processes in developing or newly industrializing countries which for economic reasons elsewhere could not be carried out or only at higher costs. So far, however, they do not suggest a strong competition between the emerging industry in the migrants' home and the existing industry in their (former) host country. Rather, the different studies prove that 'as these cross-regional collaborations multiply and deepen, both economies [in the migrants' host and home country – author's note] benefit' (Saxenian 2011: 2) due to the complementary nature of the local production processes. Even though local industries in developing countries might gradually upgrade along the value-added chain due to learning processes, the newly rising clusters there can nevertheless be regarded as spatial expansions of the existing ones resp. as their 'important partner[s] and collaborator[s]' (Saxenian 2006: 163) also in the mid- or long-term. One important reason for this aspect is the fact that the respective countries despite significant advances in science and technology still lack technological skills and research capabilities. In addition, leading edge centers like the Silicon Valley might be able to adapt to a changing business environment due to their 'ability to rapidly redeploy resources to multiple, competing experiments, and its willingness and ability to learn from failure' (Saxenian 2006: 335).

While so far a clear focus has been put on transnational entrepreneurs in high-technology sectors things might look different when discussing industries with production costs being the most important competitive factor. In such a case, the growth of new clusters in low-cost countries (developing countries or emergent markets) evoked by the activities of transnational-entrepreneurs in the long-term could rather result in the decline of established industries in developed countries with high production costs than in the co-existence of complementary cluster structures around the world. This assumption is supported by Ernst et al. (2001: 141) who in their research on industrial districts state that the 'integration into GPNs poses a fundamental change. An increased mobility of firm-specific resources and capabilities across national boundaries may erode established patterns of specialization [...]. It may also erode the strengths of existing clusters'.

Using the example of the diamond manufacturing sector, the chapter in the following analyzes whether such processes of cluster erosion can be empirically witnessed.

## Diamond Sector: An Overview

Due to significant changes like De Beers' loss of its dominant position, the debate on blood diamonds and the introduction of so-called price sheets, the diamond industry formerly widely neglected in scientific studies, increasingly has become subject to research. While most studies focus on historical and sociological aspects of the trade with diamonds (see for example Henn and Laureys 2010, Richman 2006, Henn 2010) this chapter puts is focus on the spatial dynamics of the diamond cutting industry which in the value added chain (so-called diamond pipeline) follows downstream the mining of rough diamonds, their sorting and the rough diamond trade.

The manufacturing of diamonds used to be based on manual techniques, and even though the sector has undergone a strong automatisation since the middle of the 20th century (Kinsbergen 1984), the production process still nowadays is comparatively labor intensive. In this context it also seems worth mentioning that traditionally neither cutting nor trading diamonds were based on formal trainings. Rather, the necessary knowledge used to be passed down from generation to generation and to be enriched by personal experience. Even though in the past years an increasing number of diamond dealers graduated from university and formal training courses in diamond cutting were set up, the qualification of the diamond people in the overall sector still remains on a very low level. Analyses focusing on transnational entrepreneurs in the diamond sector in this regard thus clearly differ from the above mentioned studies focusing on high-qualified 'Argonauts'.

Another central feature of the sector is that due to the enormous value of the traded goods economic transactions in the diamond sector presuppose a high level of trust between the involved parties. This was all the more important in former times with (polished) diamonds not being commoditized by standardized colors resp. other features and by selling them at prices comparable throughout the world. Not surprisingly, the diamond sector over time has developed distinct features constituting a high-trust environment like diamond bourses (i. e. large secured trading halls with strict membership regulations). Trust as a prerequisite in trading diamonds, however, is not only dependent on such a location-bound infrastructure but is also reproduced by tight social networks (Richman 2006). Especially families securing trust between a limited number of persons have evolved as an important institution in the diamond trade. According to Shor (1989: 190), families in fact must be considered as 'the glue that holds together this tradition of honor and trust. Diamond people are known by their families. In fact, it's extremely difficult to break into the business without blood connections'. The high relevance of families also reveals why the European diamond trade has been dominated by orthodox Jews who traditionally attach great value to families (Richman 2006).

Given the crucial role of implicit knowledge and trust which is necessary for carrying out economic transactions and maintained by daily face to face contacts it is not surprising that diamond trading and manufacturing used to be localized at only a handful of locations which again for several reasons have been characterized by strong micro spatial concentrations: First, security considerations still nowadays favor short ways. Second, especially the manufacturing of bigger stones requires a quick change of views between cutter (who performs contract work) and trader (who owns the stone and has information on the market resp.

pricing) since even minor cutting deviations can result in major financial losses. Third, orthodox Jewish communities having dominated the diamond business for historic and religious reasons have been characterized by their own vicinages around one or more synagogues.

Up to World War II, for historical reasons Amsterdam and Antwerp by far made up the most important diamond trading and manufacturing hubs. Apart from them, some minor cutting units also existed in France and Germany. A major spatial diversification of the sector set in with the political confusions in Europe in the 1930s. Jewish migrants who made up a large share of diamond dealers and workers mainly from Belgium migrated to the United States and Palestine thereby contributing to the emergence of new manufacturing sites in New York and Ramat Gan (Tel Aviv) (Henn and Laureys 2010). Furthermore, in 1940, the German invasion of Belgium resulted in large numbers of fugitives and, accordingly, in the emergence of new diamond centers at their escapes in Great Britain, Brazil, South Africa, Cuba, Canada, Palestine and Puerto Rico. In the aftermath of World War II, out of all these locations only New York and Ramat Gan (Israel) remained major hubs and Antwerp which had been shut down by the Nazis managed to re-emerge as the most important diamond trading center. Regardless of these developments, since the late 1960s, the world has witnessed the evolution of today's biggest diamond manufacturing cluster in India with today more than 800,000 workers while Belgium's traditional diamond industry has almost faded away (Henn and Laureys 2010).

## Methodology

For analyzing this large-scale shift in production, 65 semi-structured interviews with diamond manufacturers, diamond traders and other persons, i. e. mostly local sector experts, were carried out in Antwerp (39 interviews with diamond dealers and manufacturers, and 16 interviews with sector experts) and Mumbai/Surat (9 interviews with diamond dealers and manufacturers, and one with a sector expert) between January 2008 and April 2010.

Apart from the big DTC-sightholder companies, i. e. those companies receiving their rough diamond assortments directly from the diamond producer De Beers, diamond companies usually are small family-owned enterprises so that in most cases the managers of the companies could be interviewed. The interviews with the representatives of the companies in the first instance aimed at capturing the embeddedness of the companies in the local and global production networks, the role of families for carrying out business and the structural changes of the local diamond districts.

Given the secretive nature of the diamond business, personal recommendations (snowball system) proved to be very helpful, the more so as this forwarding of contacts complies with idiosyncratic business practices. In addition, a basic level of trust was there right from the beginning.

Typically, the interviews were conducted in the highly secured interviewees' offices; only a small number of interviews took place at the interview partners' homes or in cafés. On average, they lasted between 30 min and 1.5 hours and in most of the cases could be recorded, subsequently transcribed and analyzed.


## The Palanpuris: Some Basic Facts

While the diamond trade used to be dominated by European Jewish businessmen (Richman 2006), today dealers from India make up the most important group in terms of numbers. Many of them originated in the Western Indian town of Palanpur and accordingly call themselves Palanpuris. Besides their common origin, another major feature of the Palanpuris is their belief in Jainism, a small Indian denomination. Both their religion – as Jains the Palanpuris face certain restrictions in their choice of profession – and the fact that they belonged to the trading cast of the Banias explain why so many of them early on became specialized in commerce.

Being located in an environment of different faith the common religious background over time contributed to the merging of different joint-families to a single community which has been characterized by dense and complex patterns of kinship relations as well as by a high level of trust amongst the peers. Given the great relevance of families and trust networks in the diamond business, the Palanpuris have enjoyed significant advantages in this industry (Henn 2010): As they can make use of inherited trust in their social networks they do not have to establish trust in a timely process based on repeated interactions or to rely on recommendations by third parties amongst them. In fact, one Indian interviewee stated:

> So it's always a kind of trust [...] and faith. [...] Within the family this [trading of diamonds – author's note] can be better controlled, better handled, rather than giving it [the diamond – author's note] to somebody [...]. And this is an item [...] you can put it into pocket [...], I mean, this you easily transport, [...] and [...] [it] can disappear in a way, theoretically. So when it is within the family... – it is safety!

As was stressed in different interviews, this kind of trust will not be imposed on non-community members due to the risk that the common rules of the group would not or only partially accepted thus causing a loss of predictability of action, e. g. if a dissent occurs. From all this it becomes evident that the Palanpuris form a very close-knit community; in addition it appears that in order to activate the trust in the networks in case two Palanpuris do not know each other it is necessary to clarify the kinship relations in the wide networks before transactions will be carried out. Finally, given the fact that the trust in the network is not dependent on face-to-face interactions but only on belonging to the same community serves as an explanation why transactions amongst the peers do not presuppose spatial proximity but in principal can be carried out over large distances.

## The Palanpuris' Market Entry in Antwerp

At the beginning of the 20th century some Palanpuri jewelers moved away from their hometown to Bombay, the most important diamond and jewelry trading hub at that time. As they were successful business-wise they convinced others to join the trade (Shor 2008). 'Since this was a personalized business depending largely on trust, those whom they inducted into it were persons from Palanpur whom or whose family members they knew very well' (Dewani 1999, 35). As a consequence, the number of Palanpuri jewelers in Bombay grew rapidly. The operating distance of their businesses at that time remained restricted to parts of India and Burma. In the first decade of the 20th century, however, some Bombay-based Palanpuris had got in touch with diamond sellers from Europe, and finally started venturing to Antwerp to source polished diamonds directly (Henn 2010, Shor 2008). This, however, should not imply that the Palanpuris acted as transnational entrepreneurs at this stage since they rather carried out 'simple' import-export-businesses than making use of transnational social capital.

When World War II broke out, the early Indian engagement in the European diamond trade grinded to a complete halt. Even though the Antwerp diamond trade quickly recovered after the war (Henn and Laureys 2010), the formerly established trade patterns could not be sustained as importing diamonds to the subcontinent was forbidden when India had become independent in 1947 (Shor 2008: 27). From that moment on the Indian diamond industry thus was confined to work with the remaining outputs of some local mines resp. to recut old stones (Shor 2008). This should not hide, however, that the prewar connections to Antwerp were used by many Palanpuris for illegally 'importing' polished stones to India (Shor 2008).

When the Indian government introduced a new policy to promote diamond exports 'to earn foreign exchange and generate employment opportunities' (Dewani 1999: 35) in the 1960s, the import of rough diamonds under certain conditions became legal. In the aftermath, by falling back to their pre-war contacts to the Antwerp trade-circles, some Indian dealers began sourcing a special segment of diamonds which due to their small, irregularly shaped form was labor intensive to process and thus were rejected in Belgium and only used for industrial purposes (Henn 2010). Due to their knowledge of the production conditions at two different locations, however, the Palanpuris were aware of the fact that the low level of labor costs in India would allow for the manufacturing of these stones (later-on called 'Indian goods').[1]

---

1    Following a source dated as of 1967, the monthly income of an Indian diamond worker was about 400–1000 Belgian Francs, while almost at the same time, 8.372 Francs were announced as a minimum wage for adult diamond workers in Belgium (Henn 2010).

### Knowledge Transfer between Belgium and India

As the Palanpuris were new to the rough diamonds segment which heavily differs from the polished diamond market, gaining deeper knowledge about the industry was crucial for them. The companies in Antwerp thus not only played an important role as suppliers but also acted as important sources of knowledge. Already in the end of the 1940s, the Palanpuris had managed to win over diamond cutters from Belgium who trained some 300 cutters (Chhotalal 1990), and in 1949, the first cutting unit was set up in Surat, halfway between Palanpur and Bombay. But also later, being embedded in the Antwerp diamond networks proved to be very important for the Palanpuris for gaining knowledge of the sector. In this context, as was stressed by one interview partner recalling the 1960s as the time when the first Indians started setting up permanent buying offices in the city,

> more Indians came because we had a marvelous manufacturing business here and they came to learn the skills here what was in Antwerp at that moment. [...] At that moment, [...] an Indian person, [...] came to Antwerp. He had always five, between five or ten other people with him to learn to see how it works, to learn how the business is done here in Antwerp and dealing and sorting. [...]. They learned by the eyes. (Interview with an Antwerp-based Indian diamond dealer)

In addition, with an increasing number of manufacturing plants being established in India, the acquisition of technical knowledge became important as well. Again, Antwerp was used as source of knowledge:

> A lot of people went to Antwerp, learn with the new technology with the new tools and new and we became acquainted more and more then. (Interview with an Antwerp-based Indian diamond dealer)

Furthermore, like in the early post-war years, a transfer of knowledge also took place by engaging local cutting experts who moved to India for some years to teach the local workforce. These findings suggest that transferring the knowhow to the newly emerging center allowed for the exploitation of global differences in labor costs and thus must be considered as a major prerequisite for the subsequent take-off of the Indian industry, especially when considering the fact that the cutting expertise can easily be passed on to others. In addition, the above remarks also suggest that the newcomers from India relied on their peers already being in Antwerp in order to cope with the challenges the sector imposed on them. As such the community must be understood as an important basis for the further development of the individual businesses (Yeung 2007).

---

Generally speaking, however, it is difficult to compare production costs in diamond manufacturing due to the idiosyncratic nature of the stones.

## Market Development of 'Indian Goods'

Once having acquainted the basic knowledge about how to deal and the stones, the Palanpuris yet had to develop a market for their low-cost products. These endeavors, to a large extent were driven by external factors:

First, as Shor (2008) highlights, the US-American market for jewelry was in a state of upheaval due to the post-war boom and a restructuring of jewelry retailing. Cheap mass manufactured diamonds from India thus landed on fertile soil. The marketing of the new segment of stones by setting up subsidiaries in the US and at other locations (Dewani 1999: 40) was clearly supported by the family network of the Palanpuris. In fact, already in 1966, the first Indian diamond trader had set his foot on US-American soil, the most important market for diamond jewelry, for distributing stones polished in India. In the subsequent years, many companies followed. Also other important jewelry hubs like Hong Kong developed in these years. As a consequence and similarly to the New Argonauts (Saxenian 2006), the Palanpuris found themselves nested in structural holes able to link formerly unconnected groups of actors (i. e. buyers and sellers) (Burt 1992) in the market for cheap low-quality stones. By covering the most important nodes in the global diamond sector they started forming a global production network which combined the specific advantages of the different locations in question: Palanpuri buyers sourced rough diamonds in Antwerp and transferred them to their families in India who oversaw their low-cost manufacturing; finally the stones were transferred to relatives in the United States who sold them in New York as gateway to the world's biggest jewelry market.

Second, in the 1980s the diamond sector underwent a serious crisis. The prices for bigger, more valuable diamonds had skyrocket but only a little later fell strongly leaving a large number of manufacturers of more valuable stones insolvent. The Indian goods, however, were not affected by this development but emerged even stronger as the demand for smaller stones had increased.

Third, since 1985 the Argyle deposits in Australia have been mined. The Argyle mine not only was characterized by very rich supplies but also by the fact that the latter exactly corresponded to the specialization patterns of the Indian industry (Henn 2010, Shor 2008).

As a consequence of these developments, the market for Indian goods experienced a strong growth which prompted many Palanpuris to convince their relatives and friends to enter the business even though this can be regarded as support of competition. By doing so they acted as catalysts of business formation and thus helped the Indian diamond industry to literally take-off in the 1980s. This growth of the industry is reflected in the number of Indian diamond workers rising from about 300 in 1950 to approx. 50,000 in 1971, and finally to about 400,000 in 1984 (Henn 2010, Kinsbergen 1984). In addition, a steady expansion of the institutional basis could be observed as reflected in the establishment of the Gems and Jewellery Export Promotion Council (GJEPC) in 1966 which has dedicated itself to advisory and education services, in the foundation of the Indian Diamond

Institute (IDI) offering sector specific education and research facilities in 1971 as well as in endeavors to set up a diamond bourse in Mumbai and finally its founding in 1984 (Shor 2008).


## Increasing Cluster Competition

In the early post-war years, India for several reasons was not regarded as a major threat to the diamond sector in Belgium. First, in the 1960s with the sector employing more than 14,000 diamond workers and several thousand dealers, the small number of then only about 70-80 Indians (Henn 2010) actually did not carry any further weight. Second, different interviews suggest that the diamond people in Belgium were affected by a kind of mental lock-in: having regained the position of the world's most important center for diamond trading and manufacturing the sector was experiencing some promising years in the 1960s making the local agents neglect the developments in other regions, the more so as the Palanpuri traders bought a segment of stones which the Belgians regarded as 'rubbish' (Interview). In addition, the Indians due to their focus on stones which could not be processed in Antwerp made up a complementary market segment: The incumbent firms could make considerable gains selling these diamonds to the Palanpuris and thus had an interest in creating stable customer-supplier-relationships. In other words, both India and Belgium benefited from the activities of the new transnational entrepreneurs at this time.

However, things started to change with the first Indian dealers being appointed DTC-sightholders in the late 1960s. When De Beers had started to realize that the mentioned diamond segment now could be used for jewelry manufacturing instead of for industrial purposes only and thus be sold at much higher prices, the company increasingly supplied the emerging Indian industry with rough diamonds. As a consequence, as Belgian newspapers out of these days indicate, India more and more was regarded as a major threat especially to those companies which used to work in the smaller stones segment and were located in the outskirts of Antwerp, the so-called Kempen (Henn 2010).

In fact, only a little later the rivalry could be strongly felt in the industry: Due to the low labor costs in India, the Palanpuri dealers were able to pay higher prices for rough diamonds than their Belgian competitors. For the sake of high short-term gains, Belgian companies therefore started selling whole sights (the term refers to the high-value rough assortments of different categories of stones sold by De Beers) to the Palanpuris.

Furthermore, as a result of the growing US market and continuous technical progress the Indians received more rough diamonds, also of better categories, from De Beers while the company at the same time withdrew some segments of stones from the Belgian market. This process was accelerated by the fact that the Indians were able to gradually extend their cutting knowledge and thus to improve their technical capabilities. Indeed, the Indian import statistics suggest an increase in the average value per carat from 26 USD in 1980 to 53 USD in only eight years (Chhotalal 1990: 180). Similarly, the competition affected the segment of polished diamonds: The low labor costs enabled the Indians to reduce their prices for the stones of about 12 to 13 per cent until they reached the price level of the Belgian dealers (Tharakan 1973: 50). At the same time, an increasing number of Indian companies were able to allow their customers credits with better conditions, the more so as they enjoyed better bank lines from Belgian banks than many of their Belgian counterparts (Interview).

## Global Shift of Production

Over time, these developments brought about that the Indian diamond manufacturing sector developed to the strongest rival of the industry in Belgium (Kinsbergen 1984). For the latter, the situation due to fixed minimum wages and rigid unions became even more difficult; in fact many of the companies located in the Antwerp region could not stand the new competition and had to leave the market. Especially in the aftermath of the above mentioned sector crisis in the 1980s, the Belgian industry was severely hit by closings of companies, layoffs and a reduction of supporting infrastructure.

Of course, this does not imply that the Belgians had not reacted to the growing competition: Rather, some producers of polished diamonds set up manufacturing units abroad (especially in Thailand, Vietnam and China; No author 1997) trying to keep up with the developments but at the same time contributing to a further dismantling of the local structures. In addition, already in the 1960s it was considered to prohibit the export of rough diamonds, an idea which soon had to be rejected, however, since Belgium as a liberal market economy could not exclude single agents from buying on any local market (De Volksgazet 10/11/1967). Finally, an intensified automation 'was considered a lifeline for the high-wage manufacturing centers with the philosophy being that the increased production capacity would create a more even playing field when it came to competing with centers where the cost of labor was considerably lower' (No author 2010). According to different interview partners, for maintaining putative competitive advantages based on technologies developed in Belgium exporting cutting technologies to other countries was at first prohibited. However, as technological developments were rebuilt and also independently developed at other locations like Israel, the Indian industry with support by officials from the Argyle mine and the government (Shor 1988) from early on was able to jump on the bandwagon. Today, besides China, India in fact by far counts as the largest market for developers of diamond manufacturing technologies (No author 2010).

All in all, in the long run, the engagement of the Palanpuri entrepreneurs resulted in a global shift of the industry (Henn 2010). Figure 10.1 illustrates this development by referring to the number of Belgian diamond workers resp. to the exports of the Indian diamond industry. As can be seen, on the one hand the

**Figure 10.1   Number of Belgian diamond workers versus exports of polished diamond from India in US-Dollars (1966–2006)**

number of diamond workers in Belgium dropped from about 14,000 in 1966 to under 2,000 in 2006. Given a rising average age of the diamond polishers, a further decrease can be expected for the future. On the other hand, the Indian industry has continuously grown (2001: approx. 800,000 workers according to Henn 2010). As there are no exact time series available on the number of Indian diamond workers, however, the export value of polished diamonds which rose from 13 m USD in 1966 to 10,817 m USD in 2006 has been used as a proxy.

## Conclusions

The results of the study generally confirm that transnational entrepreneurs have a major influence on both the development of global production networks and the formation of regional clusters and thus should be taken into closer account when discussing processes of industrial transition. At the same time, however, the chapter asks for a differentiated view on the spatial implications of transnational entrepreneurship. While the New Argonauts have been found to contribute to the growth of clusters both in their home and their host regions, the results of the study suggest that processes of cluster growth in one region may induce industrial contractions in the original centers. Even though the New Argonauts distinguish themselves from the Palanpuris – the latter did not return to their home country and the majority of them are not highly qualified – this result can mainly be attributed to the sector itself. In the present case, the knowledge how to a cut a diamond almost remained unchanged over a long period of time. Due to their transnational networks, however, the Palanpuris were able to transfer the knowledge to a different location. A major competitive advantage of Belgium, the spatial concentration of competencies in diamond cutting, thus had gone. The situation intensified as the transnational entrepreneurs thanks to their economic foresight (Yeung 2007) were able to apply the newly gained knowledge in a low-cost environment for manufacturing a completely new segment of diamonds. As a long-term consequence, the Belgian cluster today is at stake.

Detached from the concrete case the findings suggest that when price competition comes to play a major role as competitive factor while knowledge creation is of minor importance, it seems probable that the knowledge transfer induced by transnational entrepreneurs can be followed by learning processes which in light of low labor costs in the migrants' home country finally might lead to an erosion of the former competitive advantage of the established locations. Even in industries having coined regional economic structures for quite a considerable time like the Belgian diamond industry, relocations and institutional dismantling appear to be probable. At this point, the question finally arises, to which degree a region will be affected by the rise of new global competitors resp. how it will respond to the growing competition from remote locations – an aspect which moves the concept closer to both policy-making and other recent approaches like regional resilience or re-bundling.

Whether the engagement of transnational entrepreneurs is regarded as contributing to regional growth or decline, however, also depends on the selected time-horizon: It could be shown that when choosing a short observation period, the Palanpuris did not compete with the incumbent firms and even must be considered as serving a complementary field of business. When applying a longer timeframe, however, it becomes apparent that they developed into the strongest rivals of the Belgian industry. Of course, such a process cannot be studied in real time; it therefore remains enthralling to see whether similar observations can be made in high-technology sectors in the future.

So far, the described erosion of clusters due to the engagement of transnational entrepreneurs may appear to be a rare phenomenon. Nevertheless, for future studies, the identification of other industries prone to or already experiencing similar developments stands out as a major task, more so as it can be assumed that due to continuous improvements in communication and transportation systems transnational entrepreneurs as elements of industrial transition will be with us for a long time to come.

## Acknowledgements

The author thanks the German Research Foundation (DFG) for financial support.

**References**

Bagwell, S. 2008. Transnational family networks and ethnic minority business development: The case of Vietnamese nail shops in the UK. *International Journal of Entrepreneurial Behaviour and Research*, 14 (6), 377–94.

Burt, R.S. 1992. *Structural Holes: The Social Structure of Competition*. Cambridge, MA: Harvard University Press.

Chhotalal, K. 1990. *Diamonds: From mines to markets*. Bombay: The Gem and Jewelry Export Promotion Council.

Coleman, J.S. 1990. *Foundations of Social Theory*, Cambridge, Mass.: Harvard University Press.

Dewani, M.D. 1999. Pioneering role of Palanpuri Jains in India's diamond industry & trade. *Diamond World*, (November-December), 35–43.

Drori, I., Honig, B. and Wright, M. 2009. Transnational entrepreneurship: An emergent field of study. *Entrepreneurship Theory and Practice*, 33(5), 1001–22.

Ernst, D., Guerrieri, P., Iammarino, S. and Pietrobielli, C. 2001. New Challenges for Industrial Clusters: Global Production Networks and Knowledge Diffusion, in *The Global Challenge to Industrial Districts. Small and Medium-sized Enterprises in Italy and Taiwan*, edited by Guerrieri, P., Iammarino, S. and C. Pietrobielli. Cheltenham: Edward Elgar, 131–44.

Henn, S. 2010. Transnational communities and regional cluster dynamics. The case of the Palanpuris in the Antwerp diamond district. *Die Erde*, 141(1–2), 127–47.

Henn, S. and Laureys, E. 2010. Bridging ruptures: The re-emergence of the Antwerp diamond district after World War II and the role of strategic action, in *Emerging Clusters*, edited by D. Fornahl, S. Henn and M.-P. Menzel. London: Edward Elgar, 74–96.

Hsing, Y. 1996. Blood, thicker than water: Interpersonal relations and Taiwanese Investments in Southern China. *Environment & Planning A*, 28(12), 2241–61.

Kinsbergen, A. 1984. *Antwerpen. Briljant aan de Top in de Diamantwereld*. Antwerpen: Provincieraad van Antwerpen.

Landolt, P., Autler, L. and Baires, S. 1999. From Hermano Lejano to Hermano Mayor: The Dialectics of Salvadoran Transnationalism. *Ethnic and Racial Studies,* 22(2), 290–315.

Light, I. 2010. Transnational Entrepreneurs in an English-Speaking World. *Die Erde* 141(1–2), 87–102.

Mustafa, M. and Chen, S. 2010. The strength of family networks in transnational immigrant entrepreneurship. *Thunderbird International Business Review*, 52(2), 97–106.

No author 1967. Bevredigende Markt maar … Angst voor India. *De Volksgazet*, 10 November.

No author 1997. Antwerp holds Steady. *New York Diamonds* (March/April), 44–45.

No author 2010. Automation brings advantages, but man has final word. [Online]. Available at: http://www.antwerpfacetsonline.be/nc/articles/single/article/automation-brings-advantages-but-man-has-final-word/ [accessed: 25 September 2011].

Portes, A., Haller, W. and Guarnizo, L. 2002. Transnational entrepreneurs: The emergence and determinants of an alternative form of immigrant economic adaptation. *American Sociological Review*, 67(2), 278–98.

Richman, B. 2006. How community institutions create economic advantage. Jewish diamond merchants in New York. *Law & Social Inquiry*, 31(2), 383–420.

RVD (Rijksverlofkas voor de Diamantnijverheid) (ed). 2006. *RVD Jaarverslag.* Antwerp: Rijksverlofkas voor de Diamantnijverheid.

Saxenian, A. 2002. Transnational Communities and the Evolution of Global Production Networks: The Cases of Taiwan, China and India. *Industry & Innovation* 9 (3), 183–202.

Saxenian, A. 2006. *The new Argonauts. Regional advantages in a global economy.* Cambridge (MA), London: Harvard University Press.

Saxenian, A. 2011. *The New Argonauts.* Diaspora Toolkit Booklet 1, Dublin: Diaspora Matters.

Shor, R. 1988. Will India be a Forge as a High-End Diamond Supplier? *JCK,* (July), 308–15.

Shor, R. 1989. Family: Diamond's Critical Connection. A look inside the tradition-bound world of diamond trading through the eyes of six families. *JCK,* (June), 190–204.

Shor, R. 2008. *The new Moghuls: The remarkable story of India's diamond people.* Unpublished manuscript. N. P.

Sternberg, R. and Müller, C. 2010. ‚New Argonauts‘ in China – Return migrants, transnational entrepreneurship and economic growth in a regional innovation system. *Die Erde,* 141(1–2), 103–25.

Tharakan, P.K.M. 1973. *India's diamond trade with Belgium: A case study in 'cross-hauling'.* Working Paper, Antwerp: University of Antwerp – Centre for Development Studies.

Yeung, H. W. 2007. Ethnic entrepreneurship and the internationalization of Chinese capitalism in Asia, in *Handbook of research on ethnic minority entrepreneurship: a co-evolutionary view on resource management,* edited by L.-P. Dana. Cheltenham: Edward Elgar, 757–98.

Chapter 11

# China as a Western Musical Instrument Producer: Case Study of the Piano Industry

Shiuh-Shen Chien, Jici Wang, Yu Ho, Mingbo Ma

## Introduction

This chapter aims to address the newly developed phenomenon that China is the biggest piano producer in the world in the 2000s. The piano was only imported into China at the very end of the 19th century. During Mao's time, the piano was regarded as a western instrument and should not be promoted, and perhaps even banned. In 1990 China produced less than 10 per cent of the world's pianos. The share of China in terms of total piano production of the world was around 30 per cent in 2000. The number dramatically increased to 50 per cent in 2003 and over 70 per cent in 2009. Total piano production in China has risen from less than 10 thousand in 1980 up to more than 300 thousand in 2009 (Figure 11.1).

China is the biggest producer not only of pianos but also of other musical instruments. For example, in 2009 China produced 930 thousand violins, which was 60 per cent of the world share. Similarly, 82 per cent of the world's wind instruments (1.4 million units), 85 per cent of guitars and other stringed instruments (5.5 million units), and 92 per cent of harmonicas (14 million units) were all made in China (China Musical Instruments Association 2010).

Chinese traditional musical instruments are categorized into eight kinds: silk (*si*), bamboo (*zhu*), wood (*mu*), stone (*shi*), metal (*jin*), clay (*tu*), gourd (*bao*), and hide (*ge*). None of them shares similar production processes and cultural acoustics with western instruments, meaning that China has limited historical linkages with western instruments (Liu and Xie 2010, Lee and Shen 1999).[1] In other words, this phenomenon deserves our academic attention: why and how and under what circumstances is China able to be the key western music instrument producer in the 2000s? This chapter, with its focus on the piano, e piano, the so-called king of western instruments due to its social popularity as well as its production complexity, aims to fill this gap.

---

1    It must be noted that western and Chinese western musical instruments are very different both in terms of production processes and cultural acoustics. For example, while both are string instruments, Chinese *erhu* and western violin are built up of different harmonic systems in relation to various preferences of intervals, melodic progression and orchestration, and temperament use.

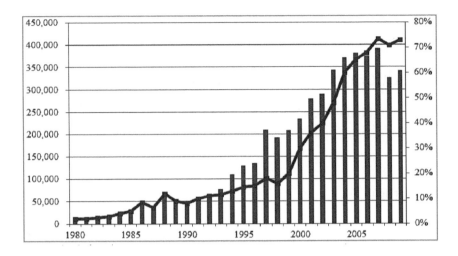

**Figure 11.1    China's piano production in the world since the 1980s**

We argue that piano production in China should be understood as a local and global production interaction in response to dramatic growth of the global piano market. These local and global production interactions under the context of post-Mao China are characterized by complicated interactions between domestic and overseas consultants, technicians, and workers from local private firms, state-owned enterprises, and/or foreign investors. Also post-Mao China 'coincidently' sees China become the world's largest piano market with more than 70 per cent of the world's pianos sold in China in 2009. This is the result of the growth of the Chinese middle class along with China's engagement with globalization. Those middle class families have more spacious houses and better incomes, but have relatively limited knowledge about music.

This chapter is divided into six sections. At the beginning, the piano production process is discussed. Second, we discuss China's piano-making history prior 1978, the year China opened up its economy and reengaged with globalization. Third, piano production in the post-Mao time is analyzed, including its overall production and sales, and production by province and by company ownership type. Fourthly, we examine several dimensions of local and global interdependencies in the production of pianos in China, such as joint investment, technology learning and so on. Fifthly, China's domestic consumption market is discussed. Finally, certain theoretical and empirical implications are noted in the conclusion.

## Process of Piano Production

Basically, a piano is a combination of strings and keyboards. The piano can be traced back to either 500 B.C., when a string instrument was invented, or the 15th century when the keyboard was created. However, it was not until the early 18th century that string and keyboard instruments could produce stable loud sounds (Xin 1999, Chanan 1988).[2] In 1709, Crisitofori, an Italian musician, invented an instrument to produce weak sounds and loud ones spontaneously through the different strengths of players' hands (Cyril 1990, Gough 1951). This is the genesis of the word 'piano', terminology shortened from the Italian word 'piano-forte'- piano (referring to weak sounds) and forte (meaning strong sounds).

Sound is undoubtedly a fundamental element of any piano. In the west, the sound of the piano is closely related to the development of classical music, which pays great attention to pitch, touch, timbre and tone. Related to that is piano production which is divided into three phrases: (1) component preparation, (2) assembly and installation, and (3) tuning and regulation (Cyril 1990, Gough 1951).

First is components' preparation. Wood is the most important material in the whole process. 85 per cent of piano materials are made of wood – including the case, cabinet, key, key-bed, key-frame, soundboard, wrest-plank, wood pillar, bass/treble bridge, action, damper and the piano stool. Different wood components are made from various sources of woods, depending on their functionalities in the piano. Wood components preparation is not just simply cutting but also drying and sealing in order to ensure no deformation under whatever atmospheric conditions the piano encounters in different geographical areas.

Another important component is the string-stretching system mainly made by steel, including the frame, strings, bridge, support blocks and brace wedges. A piano string should be both elastic enough so that it is able to be pulled tight and resilient enough in order to return back.

The action system (*ji xuan ji*), the so-called the heart of piano, is a component where the hammer strikes the string through press keys and the damper releases those keys that are pressed back freely. Tossing hammers against the string is the key mechanism for producing sounds. The functioning of the hammers and dampers is enhanced by adding felt, whose quality must be able to resist the repeated bump and still enable it to return to its initial place. Due to its complexity (for example, an action consists of 54 small part components), action of high quality is provided by a small number of professional piano component companies, one of which is located in China (see below).

The second phrase is to put different parts of the components together into a piano. For example, the string has to be put across the bridge, which is held firmly by the soundboard in order to maximize the energy from the vibrations of strings to produce sounds. The string, the tuning pins, the bridge, the soundboard are

---

2    The monochord, a sound box with a single string stretched over a movable bridge to the position, is regarded as the earliest type of piano which dates back to the Greek times (the 5th B.C.). The monochord later evolved into two kinds of instruments: psaltery and dulcimer. It was not until the 15th century that the concept of the keyboard was created and applied to stringed instruments.

installed together on the cast-iron frame. And the last stage in this phase is to put the action-keyboard system together to finish the assembly process of the piano.

The third phase is a tuning procedure in order to confirm the sound quality of any piano. Generally seven rounds of tuning are performed to make sure the length and elasticity of strings are in the appropriate condition, and to double check the pitch through tuning fork or tuner machines, and finally to adjust the voicing timbre. This phrase is very crucial as it controls melodious tones to be either clear and gentle or heavy and powerful, the fundamental characteristics of each piano. And the quality of tuning is mainly determined by professional craftsmen called tuners who have sharp ears to differentiate and adjust very subtle but critical details between professional pianos that are able to be priced high and ordinary pianos that will be sold at a lower price.[3]

To sum up, the piano production process can be divided into three phrases: (1) component preparation; (2) assembly and installation; and (3) sound tuning. While the former two are labor intensive manufacturing processes, the latter one is more cultural as it requires an understanding and appreciation of the quality of sound and music. It is noted that in the west these two aspects are both equally important. Understanding these different phrases of piano production requires us to review the recent development of China's piano industry in order to know more about whether or not China's way of piano production is similar to or different from its western counterparts. We conducted three fieldtrips in China in the first half of 2010. Data collection methods include corporate interviews, interviews with local officials, factory visits, and second-hand documentations.

## History of China's Piano Industry Development prior to 1978

In the 18th and 19th centuries, the piano was very popular for classical music performances in both Europe and the United States, where most of the pianos were produced either by music-related technicians or musicians (Gough 1951, Chanan 1988, Cyril 1990).[4] Then piano production shifted from the west to the east, first in Japan between the 1950s and the 1970s, then in Korea between the 1970s and

---

3    We appreciate the interviewees we met during our fieldtrips in China to know more about tuners. The documentary firm 'Pianomania' produced in 2008 and directed by Lilian France and Robert Cibis is also a good example to show important relationships between professional tuners and the sound quality of pianos.

4    After the first piano was created in Italy, many European branded pianos companies were established, like Bosendorfer (established in 1828), Bechstein (1853), and Petrof (1864), Perzina (1871), Schimmel (1885), Seiler (1849). In the second half of the 19th century, Britain, France, and Germany were the three largest piano producers. After the two world wars, the US replaced Europe to become the biggest piano producer in the world.

the 1990s,[5] and then in China after the 2000s (Table 11.1). As we discussed in the previous section, much work is required in the production processes of component preparation and assembly. It is the reason why spatial relocation took place from the west to Japan and Korea and then to China.

**Table 11.1    Piano production by selected countries in selected years (unit: thousand pianos)**

|  | Britain | France | Germany | America | Japan | Korea | China | The World |
|---|---|---|---|---|---|---|---|---|
| 1850 | 28 | 10 | n.a. | n.a. | n.a. | n.a. | n.a. | 40 |
| 1910 | 75 | 25 | 12 | 364 | 2 | n.a. | n.a. | 600 |
| 1950 | 40 | 20 | 20 | 170 | n.a. | n.a. | n.a. | 250 |
| 1980 | 9 | 8 | 59 | 217 | 393 | 175 | 10 | 980 |
| 1990 | 10 | 10 | 50 | 130 | 265 | 320 | 40 | 840 |
| 2000 | 8 | 10 | 33 | 100 | 90 | 145 | 235 | 780 |
| 2009 | Britain+ France+ Germany=15 | | | 7 | 30 | 30 | 324 | 490 |

*Source*: Compiled from Cyril (1990), the National Association of Musical Merchants Annual Reports (various years), China Musical Instruments Association Annual Reports (various years)

However, the piano has a very short history in China that only exists for about 150 years. The piano was first imported into China after 1842, the year China signed the Treaty of Nanking which allowed western traders and missionaries to introduce culture, materials and equipment to China. The British trader Moutries built the first piano factory in Shanghai in 1890. However, Moutries did not hire any local Chinese to work for them but employed either British or Indians who were brought by the British from outside China. It is believed that at that time Moutrie want to protect piano manufacturing skills and prevent them from diffusing to the Chinese (Wang 1997).

But Moutries soon changed its policy and hired local Chinese and trained them to make pianos. After learning certain skills, some of Chinese decided to leave Moutries and build their own factories. At the start of the 1900s, there were fourteen piano factories in Shanghai (Yang 2007). In the 1920s, establishment of piano factories also occurred in cities like Guangdong, Xiamen, Hankuo and so on.

---

5    In addition, Japanese and Korean piano companies also went overseas to expand their production. For example, Yamaha de Mexico S.A. de C.V., Yamaha's first overseas subsidiary, was established in 1958; Yamaha's US subsidiary was founded in 1960. In Europe, Yamaha built their Europe GmbH in Germany in 1966 and acquired Kemble as Yamaha's European manufacturing and market partner in 1984. Similarly, Kawai built a factory in America in 1988. See company websites: http://www.yamaha.com; http://www. piano-advice.org.uk/yamaha%20kemble%20relationship.htm;    http://www.kawaius.com/ main_links/about_us/aboutmain_09.html.

However, the piano industry in China did not develop smoothly. In the early 20th century, China suffered from World War I, World War II and its civil war, damaging critical infrastructure that had serious implications for the future development of most manufactured products, including the piano (Feng 1999, China Musical Instruments Association 2009).

After Mao's China was established in 1949, private capital and investment were not welcomed under the communist ideology. In the late 1950s, 95 per cent of piano factories were transformed into musical instrument cooperative associations and became state-owned companies (Feng 1999). For example, the Shanghai piano factory was formed by merging other local piano shops, piano factories, and component plants. There were four state-owned piano factories – Shanghai, Beijing, Liaoning (Dongbei), and Guangdong (Pearl River). The Shanghai and Guangdong factories had strong piano bases before 1949. The Beijing factory was established the capital of communist China needed some factories realted to cultural activities. The Liaoning factory was located in northeast China where there are significant forest and wood resources. The factory was established to provide cultural materials to local citizens and soldiers during the American-Korean War in the early 1950s (Liu 2009). It should be noted that other provinces also built up certain musical instruments factories, like the Xinjiang factory (stringed instruments), and the Guizhou factory (flutes).

However, the Cultural Revolution hit heavily industries that were regarded as western legacies like pianos. Playing the piano was only permitted on very limited special occasions such as the 9th National Congress of the Chinese Communist Party in 1969 and eight 'model' performances that were politically correct in terms of ideology (Liu 2009).[6] In 1976, China produced about 6,000 pianos in total, a number that was not much different from the production of 10 years earlier (China Light Industries Association 2010).

## Piano Production in Transition in Post-Mao China

After Mao's death, China opened up its economy and reengaged with globalization and made certain reforms in its musical instrument industries. In 1982, the then Ministry of Light Industry (*qing gong ye bu*) set up a musical instrument branch under the Light Industry Association of China. The association was further upgraded in 1988 to the independent Musical Instrument Industry Association directly under the Ministry of Light Industry (Liu 2009). Organizational restructuring at the ministerial level shows that the central government paid more attention to the piano industry after 1978 than it did during the Cultural Revolution. Two more dimensions should be particularly pointed out. The first is reform of state-owned

---

6   The eight 'model' performances means that those performances were designed as propaganda to promote the Cultural Revolution in the 1960s. At the time, most, if not all, western instruments were not allowed to play in such kinds of political plays.

piano factories. And the second is the promotion of private and foreign piano firms. The state-owned type took 60 per cent of China's total share in 2001, down to about 40 per cent in 2009. On the other hand, foreign ones and private ones increased from about 20 per cent in 2001 up to around 30 per cent in 2009.

*Transformation of State-Owned Piano Firms*

In the 1980s, the National Planning Council of China offered a 7 million RMB loan to reform state-owned piano factories (http:www.chinareviewnews.com, 2008/10/25). The loan was later expanded to 20 million RMB and provided significant financial resources to upgrade old facilities and machines to improve the quantity and quality of state-owned piano production. In addition, state-owned firms were allowed to improve their production performance by introducing new management programs like restructuring employee salaries and rearranging company assets (Feng 1999, Liu 2009).

The most significant example is the Guangdong Pearl River state-owned factory. In 1982, the Guangdong Pearl River abolished a life-long recruitment system and changed to a performance-based employment evaluation system, meaning that the individual salary would be highly related to the year-end performance of the factory. In addition, the factory was financed by the local authority in the 1980s and moved to a new site which was much bigger than the previous location (Feng 1999, Liu 2009). This physical expansion and institutional reform resulted in the dramatic increase of production – from around 1,000 per year in the early1980s to 24,000 in 1992, and then 70,000 in 1998. In 1999, the factory was listed as one of the big five company tax payers in the province (*bian ju bu* (Editorial Board) 1999). In 2009, the factory was even transformed to be a holding piano agglomeration, which was able to legally attract private investment from the capital market.

*Establishment of Foreign and Private Firms*

Second, institutional reforms also introduced the marketization and liberalization into the piano industry in China, meaning that four state-owned piano factories could no longer monopolize piano production. Instead, private capital and foreign investment participated in China's piano production.

For example, in 2008, there were about 30 foreign-related piano factories, both sole investments and joint investments, producing 180 thousand pianos in China (around 30 per cent of the share) (China Musical Instruments Association 2010). In terms of locations, these foreign factories were distributed in Beijing, Shanghai, Guangdong, Zhejiang, Liaoning, Jiangsu, Fujian, Shandong, Tianjin and Hubei (Table 11.2). The biggest foreign piano factory is Yamaha, followed by Hong Kong Parsons Music Corporation (China Light Industries Yearbook 2009).

**Table 11.2    Foreign-background pianos in China by capital source and by location province**

| Source Province | Hong Kong | Japan | South Korea | Other countries (incl. the US, Canada and Germany) |
|---|---|---|---|---|
| **Beijing** | Beijing Xinghai Yuehua | Beijing Kawai | — | Beijing Heintzman |
| **Guangdong** | Guangdong Zhongshan Yuehua | Guangzhou Yamaha Pearl River | — | Baldwin (Zhongshan) Zhuhai Roman Boland Chrysler Piano (China) Pearl River Kayserburg/ Ritmüller |
| **Zhejiang** | Huzhou Huapu Huzhou Huaersen Hangzhou Goodway * | Huzhou J.Sder Xiaoshan Yamaha* Hangzhou Yamaha* | — | Hua-mei Robin |
| **Shanghai** | Shanghai Haili Shanghai Artfield | — | Samick - Bechstein Piano (Shanghai). | Mendelssohn (Shanghai) Baldwin piano (Shanghai) * |
| **Liaoning** | — | Atlas (Dalian)* | — | Baldwin Dongbei (Yingkou) |
| **Other provinces (incl.Fujian, Shandong, Tianjin, Jiangsu and Hubei)** | Fuzhou Harmony Yantai Longfeng Yichang Jinbao* | Yichang Kawai | Chingdau *Sejung* Tianjin Young- chang * | Yantai Perzina Nanjing Moutries |

*Note*: * refers to those sole investment projects; others without * are joint projects between foreign capital and Chinese firms

*Source*: Compiled from Cyril (1990), the National Association of Musical Merchants Annual Reports (various years), China Musical Instruments Association Annual Reports (various years)

Yamaha again probably is the most typical case. At the beginning, Yamaha independently established Xiaoshan Musical Instrument Corporate in 1997 to produce certain piano components, like keyboards, actions, hammers and so on. In 2003, Yamaha established another independent company in Hangzhou to produce grand pianos. Xiaoshan and Hangzhou are two cities in Zhejiang province. The success of Hangzhou Yamaha encouraged Yamaha to decide to close their oversea plants in the United States and Asia (like Taiwan) and shifted its production lines to China (interview with a high-level supervisor of the Hangzhou Yamaha factory, code LS 100121). On top of that, Yamaha also built up its education center to train more people to be piano maintainers and tuners. In addition, Yamaha also established sole Yamaha sale points in many cities in China in order to promote more piano culture and consumption here.

In addition, private domestic firms also played a role in piano making in China. We discuss the case of Luoshe Township in Zhejiang province. Administratively, Luoshe is a township of Huzhou city under Zhejiang province. Luoshe Township did not have any piano history prior to 1978 but made China's first non-state-owned factory piano. By 2009, there were 65,000 pianos generated by about 50 factories in Luoshe, which was about 20 per cent of China's share (or 14 per cent of the world's share; China Daily 2010/05/04, Zhejiang *ribao* 2010/07/01).

The development of Luoshe's piano industry originated from Huzhou Piano Factory (HPF), the first non-state-owned piano factory since the establishment of communist China. Administratively, Loushe was under the management of Deqin county of Huzhou city of Zhejiang, which is a reason why the factory was named after Huzhou. HPF is very important as most piano factories in Luoshe have had certain connections with HPF- either bosses or technicians worked in HPF or some spin-off factories from HPF. For example, there were four private piano firms (Huapu, J.Sder, Louder, and Huasersen) in Luoshe producing 3,000 to 5,000 pianos in 2009. This kind of size of factories was the backbone of the China's private piano manufacture. Bosses of these four firms share one thing in common- they were then employers of HPF.

Another example is Dehua Pearl River piano company, the joint project between privately-owned Dehua firm and state-owned Guangdong Pearl River firm. DING Hongmin, the then director of Huzhou factory in the early 1990s, decided to focus more on a woodcutting factory by utilizing advantages of its wood skills in piano making. Ding co-financed with a Hong Kong company to create Dehua Wood Decorative Materials Company in 1993. Dehua has successfully upgraded from a small corporate to be a holding company which involves seven subsidiaries. Dehua Tubaobao, one of the subsidiaries targeting plywood products, was listed in the Shenzhen stock market in 2005. Ding was approached by the Guangdong Pearl River as Dehua Tubaobao was well-known for woodcutting-related work, recalling his dream of making pianos and started to think of how to 'rebuild' a piano factory (interview with a senior manager of Dehua, code LS 100122). By combining the expertise of the Guangdong Pearl River in piano making and Dehua Tubaobao in woodcutting, Dehua Pearl River became the biggest private piano producerin China. In 2008, Dehua Pearl River produced 30 thousand pianos, which contributed one third of total Pearl River Piano Group production in that year.

*Overall review on China piano production and sales after 2005*

After reviewing different contributions by various types of piano companies, we provide a review of the most recent five years of piano production and sale statistics to see the general patterns of China's piano industry development (Table 11.3). Several findings can be particularly pointed out. First, most of China's piano production is upright types, while the grand type only has less than 10 per cent. It is noted that upright pianos are technically basic and simple than grand ones. Second, China's piano export market shrank after 2008, which is believed to be related to the 2008 global financial crisis, as many western buyers decided not to purchase pianos. Different from the export market, China's piano import market grew quite dramatically even after 2008, which shows the potential of Chinese domestic consumption power. Fourthly, given that China's exports less than 20 per cent of its production to overseas, it means that most of China's production stays in the domestic market.

## Local and Global Production of Pianos in China

It is Britain's Moutrie that imported the first piano to China and trained many Chinese how to make pianos. Thus despite its limited history, China is still able to produce pianos through a complicated process of local and global production interactions. We use three examples to see how the local and global interactions have been realized in order to upgrade technology and learning in order to facilitate China's piano production (Amin 2003, Parrilli et al. 2010).

First, Luo, one of the world's leading piano's action system providers, is discussed. Luo, located in Ningbo of Zhejiang, was established in the mid-1980s as a piano component plant for the Shanghai and Guangdong state-owned firms. At that time, state-owned piano firms slowly increased their production, creating business for mature component supply firms, including Luo. In the 1990s, Luo thought about how to transform the firm, with two options – one was to produce the piano end product instead of components and parts, and the other was to be more focused on key components. Luo chose the former in the 1990s but changed to the latter strategy after the boss attended the Musikmesse Frankfurt in Germany several times in the 1990s. He was shocked about the high quality of pianos produced in the west, which – from his viewpoint – would be difficult to catch up to. In addition, he also noticed that the action system is the most valuable component in the whole piano making process (interview with a general manager of Luo, code ZJ100709). Luo started to concentrate in action making from 1992 and nowadays is the biggest action producer in the world, producing 120 thousands sets of actions by the end of 2010. Its stable provision of action systems certainly helps many of China's component and part factories to build up their assembly lines for making whole pianos.

Table 11.3  Piano production and sales in China from 2006 to 2010 (unit: thousand)

| Year | Production in China | | | China's Piano Exported to Overseas | | | China's Pianos Imported from Overseas | | | Total Piano Sales in China (4) = (1) - (2) + (3) | | |
|---|---|---|---|---|---|---|---|---|---|---|---|---|
| | Total (1) | Upright | Grand | Total (2) | Upright | Grand | Total (3) | Upright | Grand | Total (4) | Upright | Grand |
| 2006 | 375 | 343 | 32 | 114 | 91 | 23 | 21 | 18 | 3 | 282 | 271 | 11 |
| 2007 | 392 | 357 | 35 | 107 | 89 | 18 | 25 | 20 | 5 | 310 | 288 | 22 |
| 2008 | 312 | 295 | 17 | 62 | 52 | 10 | 34 | 30 | 4 | 284 | 273 | 11 |
| 2009 | 324 | 306 | 18 | 48 | 43 | 5 | 47 | 43 | 4 | 323 | 306 | 17 |
| 2010 | 382 | 361 | 21 | 59 | 53 | 6 | 67 | 63 | 4 | 390 | 369 | 19 |

Source: China Musical Instruments Yearbook (various years)

Luo's successful transformation should not only be attributed to the firm itself but also to the local and global interaction of China's piano industry. Nozaki Kinya, the first general manager of the joint firm between Yamaha and Guangdong Pearl River, was hired by Luo as a life-long consultant to provide critical knowledge in producing actions. In addition, Bechstein of Germany built up a joint technical center with Luo in 2000. Bechstein, the most expensive piano brand, also offered some technology to improve the quality of Luo's action making. In other words, without such local-global interaction, Luo might not be able to be world's leading action manufacturer.

Second, Luo's success is also related to the role of Ningboese's connection in China's piano history. Ningbo, a city about an hour by train from Shanghai, used to be famous for its woodcutting skills. As 85 per cent of a piano's raw materials are wood, woodcutting is a crucial skill in piano-making. Back in the late 1890s, Moutries asked a Ningboese woodcutter to invite his compatriots to work in piano production, which started the linkage between Ningboese and China's piano history. For example, in terms of spin-off effect, in the early 20th century, there were 14 piano factories in Shanghai. All but two were owned by Moutries, but had a Ningboese background (Wang 1997, Yang 2007). It shows the collective importance of individual Ningboese who learnt and diffused piano making technology from the Moutries.

In addition, China's first piano prior to 1949 was made by Huang Xianxing, a Ningboese who worked at Moutrie for five years. The first piano made after 1949 was produced by Wang Laian, the first director of the Beijing state-owned piano factory in the 1950s. Wang also came from Ningbo and worked in Moutries as a teenager. At the time, the other three directors of state-owned piano factories also came from Ningbo, who had more or less some connections with the Moutries or other overseas factories before they were promoted to be China's state-owned directors (Liu 2009, China Light Industries Association 2010).

This kind of individual networking facilitated the learning of China's local workers to make better and more pianos. This has also happened recently in the Beijing state-owned piano factory. In 1989, the Beijing factory hired German technicians to work as senior engineers and as vice director of the factory in order to learn more technical skills (Liu 2009). In 1990, the factory further established more institutional ways to make technology transfer from overseas possible. For example, it entered in a joint investment with Germany Heintzman to set up the Beijing Heintzman factory. In 1995 the factory signed a technical cooperation contract with Kawai, a Japanese piano company (http://www.xhpiano.com).

Third, foreign investment also actively partnered with local capital sources. Partly this was because the law did not welcome sole foreign investment and partly because overseas investors preferred to be cautious when entering China. Most foreign piano investors in the 1990s did not independently invest in China. Two strategies can be identified: either joint investment or partnering with original equipment manufacturing (OEM) firms.

For example, Harmony, the British-based piano company and China's first foreign piano investment after 1978, is a case in point. Harmony jointly established Hesheng with a Fujian company in 1985. Similarly, in 1989 Yamaha made its first step through the joint venture Tianjin Yamaha electronic musical instrument, which is nowadays the biggest electronic piano firm in the world producing 25 per cent of the world's share. In 1995, Yamaha established another joint company with the Guangdong Pearl River piano factory.[7] Different from Yamaha, Kawai, another famous Japanese piano firm, signed contracts with two local piano firms in China to do OEM for Kawai- one in Beijing and the other in Yicang (of Sichuan province). Those local and global joint investment projects certainly produced a significant quantity of pianos.

In addition, the Liaoning Dongbie factory (the official name of the company in Chinese meaning Northeast) did not operate well due to poor management. In 2004, the factory was merged with Gibson, an American musical instrument company and renamed 'Baldwin Dongbei' (Information Division, China Musical Instruments Association 2011).


## Piano Sales Market, Chinese Style

However, it is necessary to understand China's piano production is also a response to the dynamics of the global piano market. In Table 11.3 above, we showed already that most of China's piano production actually 'stays' in the domestic market instead of being exported overseas.

The rise of China's middle class certainly offers an opportunity to the China's piano industry. As noted, a piano is a luxury product. People would like to buy pianos after they get rich as a way to improve their living standards. Statistically speaking, disposable income per capita in urban China areas was raised from 1,510 RMB in 1990 to 15,781 RMB in 2008 (China Statistical Yearbook 2009). Budget spending on entertainment and cultural activities increased along with the rise of dispensable income. In addition, Chinese urban citizens used to live within a tiny room due to limited floor space. With the successful economic transition, living space per capita in China increased significantly. A big house is also important for any family to realize the idea of purchasing a piano for their children. A piano needs a spacious room.

Taken together, it is no wonder that the number of pianos per 100 households has been increasing across China as a whole- the number was only 0.5 in 1990 but increased to 1.3 in 2000, and increased to about 2.3 in 2008 (China Statistical Yearbook 2009). Some provincial-level administrations like Shanghai, Beijing

---

7   A similar story also can be found in Youngchang of Korea, which first entered China in 1993 by entering into a joint venture in Tianjin. In addition, Mendelssohn of Germany, one of the most prestigious piano producers, also established a joint venture in Shanghai in 2000.

and Shandong even had 3–5 pianos per one hundred households. And these numbers are far less than the 15 pianos per 100 households in western societies (China Musical Instruments Association 2010). Given the large population and households in China, the demand of the domestic piano market is huge, giving the China's piano production a great opportunity to develop.

However, it must be also noted that the rise of China's middle class families may be another obstacle for China's piano production, particularly in terms of quality upgrading. Despite its great quantity, China's piano quality is unable to compete with the quality of pianos produced in the west, evidenced by the price difference between the one in China and the one in Europe. For example, in terms of 9 foot (274 cm) grand piano, the European costs about 1 million RMB while the China one is about 0.2 million. A similar situation also applies to the upright piano markets (Magazine of Musical Instrument [*yue qi zi zhi*] 2007).

Two different but inter-related factors can probably explain this price variation. First, Chinese society in general does not pay much attention to musical education and the majority of people, if not to say most, do not have a high appreciation of music (Law and Ho 2009). Related to piano production is the quality and quantity of professional tuners- who play an extremely important role in adjusting pitches and trembles at the final stage of piano production. In China, the number of professional tuners until 2008 was in the hundreds (interview code LS 100125), which is obviously too small to handle the huge production that dramatically developed in the 2000s. Without appropriate tuners available to serve as the foundation of cultural embeddedness, it is understandable that producing pianos in China is just like producing other made-in-China industrial products which only focuses on component preparation and assembly. This is one reason why made-in-China pianos are not very professional compared with those produced in the west.

Second, those relatively-low sound quality pianos are 'ironically' very popular in the local piano market in China, whose major consumers are those Chinese middle class families. To buy pianos for themselves or their children is actually more or less a way to show off their social status of being middle class (interview codes LS100122, LS100124). What they demand and care about is the price instead of sound quality, which exactly fits into the supply of China's current piano production.

**Conclusion**

This chapter has answered how, why, and under what circumstances China has produced the most pianos in the world in the first decade of the 21st century. We reviewed the history of China's piano making back to the time of the 19th century. Prior to that, China had no historical linkage with the piano. In the 1900s, the piano was more or less import substitution to China. At the time, several local Chinese leared how to make pianos after they worked for Moutries, a British piano factory in China. It was only after 1978 that China emerged as a major piano producer. Nowadays, China produces 70 per cent of the world's pianos.

We argue that piano production in China should be understood in terms of its local and global production interactions in response to the dramatic growth of the global piano market. There are many complicated interactions between domestic and overseas consultants, technicians, and workers from local private firms, state-owned enterprises, and/or foreign investors. In addition, China is also a major global piano market – more than 70 per cent of the world's pianos were sold in China in 2009. There are a large number of middle class families that have emerged along with China's engagement with globalization. Those middle class families who have more spacious houses and better incomes but relatively have limited knowledge about music and care little about sound quality facilitate the development of China's piano production.

Two more points – one theoretical and one on policy – can be noted in the conclusion. The theoretical part, the 'coincidence' of China as the world's piano production factory as well as the world's piano sale market seems to reflect the industrial transition of the piano. It means that making pianos can be just like making many industrial products – focusing on factory operations without much cultural attachment.

For the policy dimension, relatively-low quality pianos produced with limited or even no cultural catchment in professional voice tuning are still accepted by the market whose main attention is about price and cost instead of pitch and tremble. This is certainly a double-edged sword- offering some opportunities for mass production (economies of scale) but creating certain obstacles for craft production (economies of scope). How state policy pays more attention to train more tuners and to promote a sense of musical appreciation to general population is probably the most important challenge for upgrading China's piano production not only qualitatively but also quantitatively in the future.

### Acknowledgement

The Authors would like to thank for financial support from the National Science Council of Taiwan and Normal Taiwan University.

### References

Amin, A. 2003. Spaces of Corporate Learning, in *Remaking the Global Economy,* edited by J. Peck and H.W.-C.Yeung. London: Sage, 114–29.

*bian ji bu* (Editorial Board) (1999) zhu jiang gang qin de zu gi (Footprints of the Pearl River Pianos). *yue qi za zhi (Musical Instruments Maganize),* 4, 6.

Chanan, M. (1988) On Science, Technology and Manufacture from Harpsichords to Yamahas. *Science as Culture,* 1, 54–91.

China Light Industries Association. 2010. *2009 zhong quo qin gong ye nian jian (2009 China Light Industries Yearbook)*. Beijing: *zhong quo qin gong ye nian jian* Publisher.

China Musical Instruments Association. 2009. *2009 zhong guo yue qi heng ye zhuan kan (2009 Special Report of China's Musical Instruments Industry)*. Beijing: China Musical Instruments Association.

China Musical Instruments Association 2010. 2010 *le qi he ye nain du bao gao* (Annual Report of China Musical Instruments Development).

Cyril, E. 1990. *The Piano: A History*. New York: Oxford University Press.

Feng, Y.-K. 1999. *xin zhong guo yue qi gong ye wu shi nian* (New China's Musical Instruments Industry over Past Fifth Years). *yue qi za zhi (Musical Instruments Maganize)*, 5, 18–21.

Gough, H. 1951. The Classical Grand Pianoforte: 1770–1830. *Proceedings of the Royal Musical Association*, 77, 41–50.

Law, W.W. and Ho, W.-C. 2009. Globalization, Values Education and School Musical Education in China. *Journal of Curriculum Studies*, 41, 501–20.

Lee, Y.-Y. and Shen, S.-Y. 1999. *Chinese Musical Instruments* Chicago: Chinese Music Society of North America.

Liu, C. 2009. *xin zhong kuo yue qi gong ye liu shi nian xuan li* (Review on Sixty Years of China's Musical Instruments Industry). *yue qi za zhi (Musical Instruments Maganize)*, 10, 6–11.

Liu, J. and Xie, L. 2010. Comparison of Performance in Automatic Classification between chinese and Western Musical Instruments. In *2010 WASE International Conference on information Engineering (ICIE)*, 3–6.

Parrilli, M.D., Aranguren, M.J. and Larrea, M. 2010. The Role of Interactive Learning to Close the 'Innovation Gap' in SME-Based Local Economies: A Furniture Cluster in the Basque Country and its Key Policy Implications. *European Planning Studies*, 18, 351–70.

Wang, D. 1997. *shanghai er qin gong ye zhi [The Annals of Shanghai Second Light Industries]* Shanghai: Shanghai *shehui kexue yuan*.

Chapter 12

# Spatial and Organisational Transition of an East Asian High-Growth Region: The Electronics Industry in the Greater Pearl River Delta

Daniel Schiller

## Introduction

The transition of industrial spaces and organisation in emerging regions of East Asia differs markedly from the Western experience. Among other things, high growth rates, close interrelations with the global economy and a comparatively high speed of institutional change have resulted in accelerated processes of structural change and increased manufacturing flexibility. Consequently, Park (2000), Yeung and Lin (2003), and Yeung (2006) outlined a research agenda for an economic geography of Asia, while Li and Peng (2008) and Barney and Zhang (2009) identified future trajectories for management research on China. This chapter picks up some of the suggested conceptualisations of spatial organisation within Asian regions and combines the spatial focus of economic geography with the firm perspective in management science.

The local impact of global production networks (GPNs) has been a major topic in the recent debate among economic geographers working on Asia (Yeung 2009). This literature has provided powerful evidence for the global-local dynamics of industrial transitions. However, it has been largely focused on the 'global' aspect as the major driving force of local configurations, while it has partly overlooked the importance of specific spatio-institutional contexts and endogenous domestic dynamics. While it goes without saying that the growth of Asian regions was induced by their integration within the global division of labour, domestic demand and indigenous firms have become an independent factor more recently, especially in the case of China and in the aftermath of the global economic and financial crisis.

A perspective on the internal organisation of regional growth models in China has been put forward by Wei et al. (2007, 2009) and Luo and Wei (2009), who studied the idiosyncrasies of regions such as Wenzhou, Suzhou, and Nanjing. Naturally, this approach did not cover the global perspective as comprehensively as the GPN studies. By taking both strands of research into account, this chapter positions itself at the interface of global-local dynamics of GPNs and the interrelated spatial and organisational transition within a specific region.

The electronics industry of the Greater Pearl River Delta (GPRD) in southern China was chosen here as a regional-sectoral case study. It comprises Hong Kong (HK) and nine cities of the Guangdong Province, also known as the Pearl River Delta (PRD). This spatial setting contains two important institutional features: firstly, cross-border relations between mainland cities and the HK Special Administrative Region (SAR), and secondly, the ongoing institutional transition within mainland China itself.

The GPRD has been among the fastest growing regions in the world during the last three decades and has become a major cluster of production in many industries (Enright et al. 2005). Today, the GPRD region is one of the major global production hubs of the electronics industry, which is by far its most important manufacturing sector in terms of value creation, foreign investment and trade. The specific organisation of large parts of the electronics industry within a modular value chain (Sturgeon 2002) results in a set of complex global-local dynamics among leading firms, contract manufacturers and parts suppliers.

High growth rates and institutional change have resulted in extensive spatial and organisational transitions within the GPRD. Firms located in cities that were industrialised in the 1980s, particularly in Shenzhen, added new functions besides production, while new production sites emerged in other parts of the region, particularly in the western PRD. Early investors from Hong Kong have been supplemented by other foreign investors (especially from Taiwan and Korea) and newly founded private domestic firms. Institutional change and competitive dynamics have resulted in organisational transitions, such as new forms of corporate governance as well as inter-firm and government relations.

It is the aim of this chapter to describe and to provide some explanations for recent spatial and organisational transitions in the electronics industry of the GPRD based on data from two company surveys and qualitative interviews of electronics firms in HK and the PRD. The following central research questions will be used to structure the conceptual and empirical sections:

*Spatial Transition*

- Which positions in the global value chain of the electronics industry are held by the surveyed firms from HK and the PRD?
- How are the surveyed electronics firms organised spatially within the GPRD and how did these patterns evolve over time?

*Organisational Transition*

- What kind of governance modes do the surveyed electronics firms from HK apply to their cross-border production activities in the PRD?

The chapter is organised in five sections, discussing conceptual underpinnings adapted to the Chinese context, introducing the data and methods used for empirical analysis, presenting results on issues of spatial and organisational transition, and deriving some conclusions on policy recommendations and further research perspectives.

## Conceptualising Spatial and Organisational Transition in High-Growth Regions

*Outsourcing and Offshoring in Modular Value Chains of the Electronics Industry*

Pioneered by American electronics firms, the industry has undergone a double transition of outsourcing and offshoring and a comprehensive overhaul of its value chain organisation (Langlois 2003). Starting from the early 1990s, large electronics manufacturers, such as IBM and Apple, sold their manufacturing facilities to a new species of contract manufacturers (CMs) that were subsequently offering electronics manufacturing services (EMS) to several leading firms, i.e. the global brand owners. The term 'modularisation' was introduced by Sturgeon (2002) to capture this transition, which was later applied by electronics firms from Europe and Asia, particularly from Taiwan. Only a small number electronics firms, especially those from Japan and Korea, maintained a rather captive organisation of their manufacturing.

The disentangled modular value chain paved the way for a new wave of offshoring standardised manufacturing activities to locations with more favourable factor endowments, especially manual labour, and the emergence of global production networks (GPNs, Ernst 2005). The modular organisation of GPNs provided entry points for other firms, especially parts suppliers, to the electronics value chain. While contract manufacturers set up production facilities in Asian regions or had even originated from countries such as Taiwan, Korea, Singapore, or Hong Kong, parts suppliers of foreign and domestic origin also mushroomed in these locations and, eventually, an Asian electronics industry was created (Mathews and Cho 2000).

The major mechanism of technological and organisational upgrading was described by Hobday (1995) as the transition from original equipment manufacturing (OEM) via own-design manufacturing (ODM) to own-brand manufacturing (OBM). During this transition, latecomer firms acquire additional technological capabilities, such as design of products, production lines, marketing and technological innovation over and above initial capabilities, which allows them to operate a pre-installed production system according to the specifications of the customer. Notwithstanding the fact that the electronics industry arrived in Asia before the emergence of modular production models, and that considerable efforts to upgrade the technological competences of national champions had been made, strategies of supplier-oriented upgrading within global production networks provided more flexible and less resource-consuming entry points for latecomer firms (Sturgeon and Lester 2003). Based on these considerations, the empirical analysis will provide evidence pertaining to the position of firms from HK and the PRD within the global electronics value chain.

*Evolution of Localised Production Systems*

Localised production systems have been described by a large number of different, but partly overlapping concepts such as industrial districts, clusters, new industrial spaces; see Moulaert and Sekia (2003) for a comprehensive review of what Markusen (1999) called 'fuzzy concepts'. These models will not be reiterated here, but illustrative applications that are relevant to the transition within high-growth regions will be discussed. In general, high-growth regions in East Asia are defined preliminarily as regions that have realised outstanding economic growth and industrialisation at an unprecedented speed based on their export orientation and, more recently, domestic markets (Yeung 2009). Within China, the PRD and the Yangtze River Delta (YRD) both fit with this definition. Two approaches which apply concepts of localised production systems directly to East Asian regions and which have added a dynamic perspective on spatial transitions will be discussed briefly in the following section: (i) the evolution of industrial districts introduced by Markusen (1996) and adapted to the context of Asia by Park (1996) and Guerrieri and Pietrobelli (2004), and (ii) the strategic coupling approach by Yeung (2009).

According to Markusen's (1996) typology of industrial districts, the PRD could be interpreted as a satellite platform district which is dominated by large, externally owned and headquartered firms (Yang 2009), or partly as a global production network as outlined by Guerrieri and Pietrobelli (2004), in which the local production sites are domestic small and medium sized enterprises (SMEs) instead of affiliates of global firms, but closely related via original equipment manufacturing (OEM) relationships. Its most important feature is the weakness of connections within the region and the predominance of links to the parent corporation and branch plants in other regions. By comparing the embeddedness of Taiwanese PC firms in the PRD and the YRD, Yang (2009) has shown that the PRD model is characterised by few initiatives to promote strategic coupling between local and foreign manufacturers. Referring to the case of Shenzhen within the PRD, Park (1996) noted that its transition from a new satellite industrial district developed from foreign investment towards one with more local firms and networks began during the 1990s. Later, Lu and Wei (2007) found some evidence that the mechanisms of regional growth in the PRD are becoming increasingly influenced by a regionally integrated development rather than by external factors.

In general, the strong manufacturing base of electronics firms in the PRD is a viable starting point from which to transform the platform-like system into a regionally coherent production and innovation system. Several clusters of the electronics industry have been identified in previous studies of the region (Bellandi and di Tommaso 2005, Enright et al. 2005, Li & Fung Research Centre

2006), but most of them are spatial concentrations of manual labour-intensive activities which co-locate because of a large pool of low-skilled migrant workers, rather than because of the opportunity for intensive inter-firm collaboration and knowledge exchange (Wang and Mei 2009).

By looking at the transition of localised production systems in East Asia from the global perspective, Yeung (2009) analyses the modes of strategic coupling between local players and leading firms in GPNs. He identifies three key dynamics of regional development trajectories: strategic coupling (i) through international partnerships, (ii) through indigenous innovation, and (iii) through the provision of production platforms. Even though it is acknowledged that more than one type might apply to a particular region, the taxonomy allows a distinction to be made between different global-local configurations. Within this framework, the PRD provides more of a production platform, but with some evidence of a transition towards indigenous innovation by leading domestic private firms, especially in the field of electronics.

*Organisational Transition of Governance Modes*

Li and Peng (2008) highlighted the need for hybrid theories of the Chinese firm that take into account systematic differences in the assumptions behind Western and Chinese theories. The examples of recent major streams of management research in China provided in their commentary are closely related to the governance of business organisations in China and the role of social relationships and connections.

Three basic modes of governing economic activities are set out in the relevant literature: market, hierarchy, and hybrid forms, i.e. networks (Powell 1990). Within the modular value chain, the modular network is a specific mode of governance, which differs from Powell's relational network due to less relational, but more price-based communication and less mutual dependency (Sturgeon 2002). However, due to the emergence of large contract manufacturers, the dependence of leading firms on their producers is also remarkable in many cases. Wang and Nicholas (2007) differentiate between networked forms of governance by analysing equity relationships. In the case of China, contractual non-equity joint ventures are distinguished from those in which both partners hold equity.

Transaction cost economics (Williamson 1998) discusses the determinants that decide about governance modes which are shaped by specific characteristics of the transaction, i.e. asset specificity and frequency. In addition, the institutional environment determines the degree of uncertainty, since it constrains human (opportunistic) behaviour, with an efficient and reliable institutional setting thus decreasing uncertainty (Meyer et al. 2009).

Within a modular value chain, it is expected that the borders of the firm are defined by the degree of codifiability of knowledge regarding product and process specifications (Sturgeon 2002). The leading firms transfer codified specifications about the final product to their contract manufacturers. They, in turn, will break up the product into components which they might source from the market if the products are less specific, as is often the case with basic electronics components. Therefore, a high degree of market relations is expected at the lower end of the electronics value chain, and therefore in emerging regions. However, since the institutional environment and related uncertainty interfere with the technical determinants, it is expected that a transition of governance modes from intra-firm transactions within a hierarchy of affiliated companies towards market-based or networked modes takes place over time in regions characterised by a high degree of institutional change like in China.

These general trends of spatial and organisational change in the electronics industry are most often explained by looking at the lead firms of the industry in the USA, Japan, Taiwan or Korea. But it is often neglected to observe whether and how particular local institutional settings result in differences of spatial and organisational strategies. This chapter therefore aims at focussing particularly on the hybrid institutions that are in place in China and in the cross-border setting of HK and the PRD. An integrative and dynamic perspective is used to add context-specific insights on value chain upgrading, spatial clustering, and firm organisation and governance to the existing literature.

## Methodology

The empirical part of this chapter is mainly based on two company surveys that were carried out in HK and the PRD separately during 2007/2008. Additional information is drawn from in-depth interviews with ten large HK electronics firms and other studies of transitions within the regions, especially those of HKPC (2003), Enright et al. (2005), FHKI (2007) and Hürtgen et al. (2009).

In HK, a survey of SMEs within the electronics industry was conducted in the second half of 2007. Companies with less than 100 employees in HK are the most important group of electronics firms on the spot, but they normally possess much larger production sites in the PRD or elsewhere in China. They were chosen as the survey population since their operations are easier to cover by a standardised method, while large firms and conglomerates were interviewed qualitatively. The sampling of firms was based on the company database of the Hong Kong Trade Development Council (HKTDC). From 4,903 registered Hong Kong SMEs in the electronics industry, a random sample of 2,000 firms was contacted. Finally, 104 CEOs or senior managers were interviewed face-to-face using a standardised questionnaire. The average duration of the interviews was approximately one hour. Since the collection of empirical data on the influence of institutions on governance modes is still at an explorative stage, an attempt has not been made to produce representative information. The interviews instead concentrated on the strategic behaviour of Hong Kong headquarters, the firms' international activities and their linkages to the PRD.

In the PRD, the set of primary data covers 222 electronics firms. The survey was focused on the cities of Dongguan (89 firms or 40 per cent of the sample) and Guangzhou (116 firms or 52 per cent). The remaining 17 firms or 8 per cent are located in other cities of the delta. Guangzhou was selected as one of the cities in which a high portion of Guangdong's electronics industry is located (7 per cent of the provincial gross value added in electronics). It is the provincial capital with a diversified industrial structure, a strong service sector, and is home to the major universities of the province. Firms in Guangzhou are markedly focused on the domestic market. In contrast, Dongguan was initially industrialised using investments from Hong Kong and Taiwan (Yang 2007). Its economic structure is almost completely focused on the manufacturing sector, and in particular on the electronics industry. 12 per cent of the gross value added of the electronics industry in Guangdong is produced in Dongguan.

The sampling of firms for Dongguan was based on the Guangdong Electronics Company Catalogue 2007 and for Guangzhou on a list provided by the Statistical Bureau of Guangzhou. A sampling according to firms' characteristics such as size was not possible, since additional information on the firms was missing in the sampling frames used in the PRD. The survey was focused on districts with a high density of electronics companies. In Guangzhou, the Tianhe, Panyu and Huadu districts were selected as relevant clusters of the electronics industry. In Dongguan, the three main districts of the survey were Changan, Dongcheng, and Houjie. 300 questionnaires were distributed to randomly selected firms in these districts. In Dongguan, 89 of 150 questionnaires, or 59 per cent, were returned. The response rate in Guangzhou was 77 per cent, or 116 of the 150 distributed questionnaires. All questionnaires could be used for the analysis, as missing or contradictory answers were redressed in telephone follow-ups. The survey was completed in early 2008. An overview of some characteristics of the interviewed firms is given in Table 12.1.

**Table 12.1   General characteristics of the surveyed firms in HK and the PRD**

|  |  | HK | PRD | Dongguan | Guang-zhou |
|---|---|---|---|---|---|
| Number of firms |  | 104 | 222 | 91 | 116 |
| Ownership in % | Chinese | 3% | 53% | 40% | 61% |
|  | HK | 89% | 18% | 15% | 21% |
|  | Taiwan | 3% | 12% | 22% | 5% |
|  | Other | 5% | 17% | 23% | 13% |
| Employees in PRD in % | 1 – 99 | 46% | 36% | 27% | 43% |
|  | 100 – 499 | 28% | 38% | 45% | 34% |
|  | 500 and above | 26% | 26% | 29% | 23% |
| Foundation in PRD in % | 2000 and earlier | 75% | 58% | 60% | 56% |
|  | 2001–2007 | 25% | 42% | 40% | 44% |

*Source*: own survey

Right after the survey, the region was strongly affected by the global economic and financial crisis which definitely had a short-term impact on the surveyed firms and the transition of the region. However, the data and results presented here are expected to remain valid for two reasons. First, despite its strong export orientation, the electronics industry in the GPRD recovered very quickly and exceeded its pre-crisis level of economic activity already in 2010. Second, interviews in the post-crisis period confirmed that ongoing long-term structural changes, which are the focus of this chapter, were accelerated by the crisis, but new growth paths were not nurtured during this short period of economic turmoil.

## Empirical Results on Spatial Transitions

*Positioning the Greater Pearl River Delta within the Global Electronics Value Chain*

Electronics firms within the GPRD entered the global electronics value chain by supplying parts and components to contract manufacturers or by producing goods under OEM arrangements without their own designs or direct marketing to final consumers (Hürtgen et al. 2009). In 2007, 77 per cent of electronics firms in HK surveyed by the HK Trade and Development Council were still engaged in OEM activities, while 70 per cent performed ODM and 49 per cent OBM activities. The figures were about the same as those in 2003 (HKTDC 2008). However, the ODM business is becoming more important for HK manufacturers in terms of sales due to rising labour costs in the PRD, which is the main manufacturing base of HK firms, and due to keen competition from other locations. The survey data from 2007 are very much in line with these findings (Table 12.2). Firms from HK are shifting away from pure OEM activities (only 18 per cent of firms founded in 2001 or later) and complement them with ODM and OBM (41 per cent). However, a comparatively smaller group of firms (23 per cent) have given up OEM activities completely.

**Table 12.2    Sales profiles of electronics firms in HK and the PRD according to age of the firm**

|  | Firms in HK | | Firms in the PRD | |
|---|---|---|---|---|
|  | 2000 and earlier | 2001 and later | 2000 and earlier | 2001 and later |
| **Main Location of Sales** | | | | |
| Industrialised Countries | 56% | 78% | 30% | 24% |
| Emerging Asia | 28% | 13% | 20% | 14% |
| Mainland China | 16% | 9% | 50% | 61% |
| chi-test sign. | 0.000 | | 0.048 | |
| **Main Sales Categories*** | | | | |
| Purely OEM | 31% | 18% | 28% | 23% |
| Mainly OEM | 35% | 41% | 28% | 17% |
| Mainly ODM/OBM | 20% | 18% | 20% | 37% |
| Purely ODM/OBM | 14% | 23% | 23% | 23% |
| chi-test sign. | 0.019 | | 0.007 | |
| **HK Customers** | | | | |
| Yes | ... | ... | 48% | 27% |
| No | ... | ... | 52% | 73% |
| chi-test sign. | ... | | 0.000 | |

*Note*: *Within the PRD, only for firms with main location of sales outside Mainland China
*Source*: own survey

Managers of large electronics firms in HK mentioned in personal interviews that they are still using the OEM business to maintain a stable basic throughput of products on their assembly lines. At the same time, they are using their manufacturing experience as a stepping stone for developing their own designs and brands. However, even some of the largest electronics firms in HK with strong brands are still involved in OEM to a certain degree. A major reason for this is the fact that developing one's own brand is not only time consuming, but also risky. One recent example is that of a large battery producer in HK that successfully introduced products under its own brand, but became involved in a serious case of cadmium poisoning at major production sites in the PRD. As a consequence, they had to scale down their OBM business, but were able to survive because of their OEM sales. Nevertheless, several managers of large HK electronics firms acknowledged that the prospects for introducing HK brands in China and other emerging markets are very good because they have a comparatively good reputation at a competitive price (see also HKTDC 2008).

These qualitative insights show that a clear distinction between OEM, ODM and OBM production occurs very rarely in reality. As a consequence of these findings, the measurement of production modes should incorporate a dynamic view of the evolution of production strategies which potentially offers further insights into the transition from OEM to more demanding production modes.

Main sales categories were analysed for firms located in the PRD. However, only firms with foreign owners (including Hong Kong and Taiwan) were selected due to the differences between mainland Chinese and foreign firms. Chinese firms are, in many cases, oriented towards the domestic markets, and introducing their own design or brand within China is expected to be easier than in the international arena. Therefore, the percentage of domestic Chinese firms purely carrying out ODM/OBM is very high (41 per cent), which is obviously not comparable to the shares among foreign firms (23 per cent). This example shows that data regarding sales categories have to be interpreted by looking at the sophistication of the markets that are penetrated by these sales in order to derive a conclusion about technological or marketing capabilities.

Foreign firms located in the PRD have moved away from pure OEM activities over time (Table 12.1), but younger firms are less often focused solely on the more risky ODM/OBM business than older firms. This provides further evidence for the conclusion that particularly younger firms with fewer resources need their OEM business as a basis for introducing more sophisticated business models.

Further information on the main markets for sales is provided in the lower part of Table 12.1, which depicts diverging paths of transition. Younger firms in HK are oriented to an even larger degree towards industrialised countries, while the importance of sales to China and other emerging countries in Asia is decreasing. This is obviously a result of the fact that firms from the PRD are increasingly able to organise these sales activities on their own. While the integration within the electronics cluster of the PRD is accelerating quickly, an even larger share of sales by PRD firms is directed towards domestic customers. At the same time, HK's role as a broker for products from the PRD is decreasing in relative terms. Every second firm among the older ones still has a customer in HK, but this share decreases to every fourth firm among the younger ones. However, among those firms located in the PRD that are selling products to customers abroad, especially to markets in industrialised countries, the share of firms having a customer in HK is still very high. Therefore, HK's relative decline in importance for firms from a more domestically oriented PRD does not necessarily result in a limited importance of HK for the internationalisation of Chinese firms in general.

*Evolution of Spatial Patterns within the GPRD*

Rapid spatial transition was the dominant feature within the GPRD region after the beginning of economic reforms in China from 1978. In the early days, HK manufacturers relocated their production facilities to the PRD and established a production model that was called 'front shop, back factory'. Over time, this relationship has been transformed and enhanced to a remarkable degree. On the one hand, simple production functions of firms in the PRD have been upgraded through additional service functions in fields such as sales, finance and innovation, which has created a more interactive and cooperative partnership between HK and the PRD. On the other hand, the dominant mode of entering the PRD via HK has been complemented by foreign firms investing directly in the PRD without setting up offices in HK, and by endogenous domestic firms. Within the PRD, rapid spatial shifts of economic activities and strong clustering effects have been observed (Enright et al. 2005, Li & Fung Research Centre 2006).

*Modes of entry to the PRD*
The cross-border 'front shop, back factory' model between HK and the PRD became the major spatial feature of production organisation within the GPRD. In the past, more than 70 per cent of all investments in the PRD were channelled through HK, and HK's share of foreign investments in the PRD is still around 50 per cent now (HKCPU 2010).

   Within the Guangdong Province, foreign-owned firms are as important in terms of gross value added in the industrial sector as domestic firms (Table 12.3). Within the electronics sector, their share is even above 60 per cent, while state-owned enterprises are rather unimportant in this sector. As indicated by the share of HK in total FDI, companies owned by Hong Kong, Macau, and Taiwan (HMT) are responsible for more than 50 per cent of the foreign-owned industrial value added. The two survey sites differ in terms of ownership: Dongguan is very strongly focused on foreign firms, especially those from HMT, while a larger share of firms in Guangzhou is domestically owned. Within the foreign-owned sector of Guangzhou, other foreign firms (not from HMT) dominate.

**Table 12.3    Share of industrial gross value added according to ownership**

| Ownership | All firms in Guangdong | Electronics firms in Guangdong | All firms in Dongguan | All firms in Guangzhou |
|---|---|---|---|---|
| State-owned | 18% | 4% | 10% | 27% |
| Private domestic | 27% | 33% | 12% | 16% |
| Foreign-invested | 53% | 62% | 77% | 56% |
| thereof HMT* | 29% | … | 44% | 19% |
| thereof others | 24% | … | 33% | 37% |

*Note*: *HMT - Hong Kong, Macau, and Taiwan

*Source*: own calculations based on Guangdong Statistical Yearbook 2008

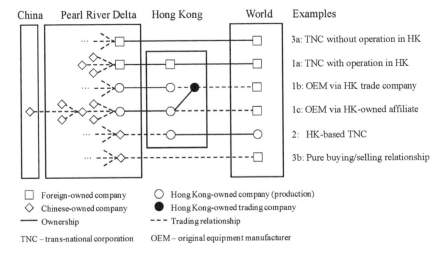

**Figure 12.1    Modes of cross-border production organisation in HK and the PRD**

Compound investment data does not include any information on the organisational mode of entry to the PRD. Therefore, Figure 12.1 attempts to differentiate between possible modes of entry. The figure only covers those production modes that are oriented towards the international market, but excludes Chinese firms that are producing for the domestic market. In general, a foreign firm that wishes to enter the PRD has to decide whether to use HK as an entry point or not. If HK is included in the global value chain, this might happen via an office set-up by the foreign firm which is used to control production activities in the PRD (Example 1a in Figure 12.1). However, leading global firms in the electronics industry often contract HK-owned firms to organise the production in the PRD. This is organised either via a trading company that is sourcing products from the PRD (1b) or via an OEM arrangement (1c). As shown above, several HK-owned firms themselves became organisers of global value chains over time and set up bases abroad (2). If leading global firms are entering the PRD directly, they can either set up their own production site there (3a) or enter into a buying/selling relationship with a local firm without any equity relationship (3b).

The quantitative importance of different modes of entry to the PRD by foreign firms and the relevance of the domestic sector is assessed by looking at the role of connections to HK for different kinds of ownership (Table 12.4). It can be seen that a connection to HK (examples 1a–c in Figure 12.1), regardless of its size and nature, still exists for most foreign-owned firms and even for half of the Chinese-owned firms. The nature of this connection is further subdivided in Table 12.5. In addition, about one-third of all foreign-owned companies indicated that they use HK as their most important sales platform. HK is chosen as the main sales channel almost as often as direct sales to overseas markets (examples 3a–b in Figure 12.1). However, the figures in Table 12.4 also provide evidence for the ongoing spatial transition of the GPRD because several alternative models have emerged besides the traditional HK-PRD cross-border production model. The domestic market has become the major destination for most Chinese-owned firms, and even foreign firms are selling a considerable amount of their products locally. At the same time, HK is not only a broker for sales of final products to industrialised countries, but has also maintained an important position in those parts of the electronics value chains that are localised in emerging Asian economies.

**Table 12.4    Spatial sales patterns of firms in the PRD according to ownership**

| Main sales channel | Ownership | | | | Total |
| --- | --- | --- | --- | --- | --- |
| | Chinese | Hong Kong | Taiwan | Other | |
| Connection to HK (sales and others) | 48% | 100% | 73% | 73% | 65% |
| Mainly sales via HK to... | 10% | 44% | 31% | 30% | 22% |
| ... industrialised Countries | 4% | 15% | 15% | 24% | 10% |
| ... emerging Asia | 4% | 21% | 12% | 7% | 8% |
| ... mainland China | 3% | 8% | 4% | 0% | 3% |
| Mainly sales to overseas markets | 20% | 28% | 50% | 39% | 28% |
| Mainly sales to the domestic market | 70% | 28% | 19% | 30% | 50% |

*Source*: own survey

*The importance of linkages to HK*
Even though firms are more frequently entering the PRD without setting up a basis in HK, the majority of electronics firms in the PRD remain connected in one way or another with HK. In total, two-thirds of all electronics firms surveyed in the PRD still maintain some kind of business ties with HK, for example by selling or sourcing products via HK, ownership relations, or financing. However, the number of firms that indicated maintaining any kind of linkage or connection with HK has significantly decreased, which becomes clear when older and younger firms are compared (Table 12.5).

**Table 12.5**   **HK connections of electronics firms in the PRD according to age of the firm**

| Type of HK connection | 2000 and earlier | 2001 and later | Chi-square-sign. |
|---|---|---|---|
| Any connection to HK | 73% | 53% | 0.002 |
| Mainly HK-owned | 25% | 8% | 0.001 |
| Main customer in HK | 48% | 27% | 0.002 |
| Sales via HK affiliate | 24% | 12% | 0.025 |
| Sales via HK trading company | 16% | 11% | 0.276 |
| Sales via other HK company | 9% | 10% | 0.810 |
| Sourcing via HK | 36% | 17% | 0.003 |
| Innovation in HK | 5% | 2% | 0.309 |

*Source*: own survey

Table 12.5 also reveals clear differences in importance between the types of HK connection and their decline in importance over time. In line with the findings from the official statistics above, the proportion of mainly HK-owned firms is much lower among younger firms. Customer-producer relations have been the most important link between HK and the PRD in the past, and they still are today, but at a much lower level. Sales via HK affiliates in particular have become less important, while the reduction of sales via trading companies is statistically not significant. Sourcing is even more often carried out without the involvement of HK. In the field of innovation, HK was not able to transform its comparative strengths in terms of skills and finance to improve its negligible role for firms in the PRD.

Further insights into the spatial division of labour in the GPRD are gained from the main locations of four different business activities chosen by firms in HK and the PRD, as set out in Table 12.6. The table shows that the PRD has become a very integrated production cluster in which the majority of business activities are carried out locally even by foreign firms and often without the direct involvement of HK. While the almost complete outsourcing of production activities to the PRD is a well-known fact, the figures provide evidence that the PRD has become the main location of most firms for several other activities that were widely deemed to be comparative advantages of HK in the past.

**Table 12.6    Main location of business activities by electronics firms in HK and the PRD**

| | HK | | PRD (only operation, multiple answers possible) | | | |
|---|---|---|---|---|---|---|
| | Operation | Decision | Chinese | Hong Kong | Taiwanese | Other |
| **Production** | | | | | | |
| HK | 8% | 49% | 1% | 6% | 5% | 0% |
| PRD | 90% | 49% | 96% | 100% | 100% | 97% |
| Other | 1% | 2% | 5% | 0% | 0% | 3% |
| **Sales/ Marketing** | | | | | | |
| HK | 84% | 84% | 8% | 55% | 30% | 36% |
| PRD | 15% | 14% | 89% | 67% | 57% | 61% |
| Other | 1% | 2% | 13% | 3% | 17% | 39% |
| **Finance** | | | | | | |
| HK | 83% | 87% | 5% | 37% | 5% | 21% |
| PRD | 16% | 10% | 91% | 83% | 73% | 79% |
| Other | 1% | 4% | 4% | 0% | 27% | 10% |
| **Innovation** | | | | | | |
| HK | 40% | 66% | 0% | 18% | 5% | 5% |
| PRD | 59% | 32% | 93% | 94% | 86% | 86% |
| Other | 1% | 2% | 9% | 0% | 29% | 27% |

*Source*: own survey

Sales and marketing activities are still the second most important task that is carried out via HK by firms located in the PRD, but domestic Chinese firms do not connect with partners in HK for that task, and even a majority of HK-owned firms indicated their mainland site to be the main location for sales and marketing activities. The financing of business operations is a traditional strength of HK as a global financial centre. However, the data reveal that the mainland is now offering better opportunities to finance business operations directly via domestic banks, via HK and via some foreign banks that have moved to the region recently. With regard to innovation, the figures in Table 6 confirm the supposition made above that HK has not become a major provider for new technologies for the manufacturing sector in the PRD despite its strong universities and highly-skilled workforce. Even the majority of firms surveyed in HK indicated that the PRD is their main location for carrying out innovation activities.

The results are supported by an earlier qualitative assessment of locational choices by electronics firms from different countries in the GPRD (Enright et al. 2005). He pointed out that HK is only chosen as a location by HK and Taiwanese-owned firms. Those operations which are carried out relatively often in HK by these firms include management, finance, sales and marketing, and logistics, while the majority of other tasks are relocated to the PRD.

The qualitative interviews provided four sets of explanations for the changing division of labour between HK and the PRD:

1. Increasing capabilities of local firms and skills of employees as well as increasing advantages from dense local production clusters.
2. Improved institutions that reduce uncertainties for the performance of advanced business functions, though this improvement was assessed to be rather slow.
3. A pull by market forces due to the increasing local demand and emerging lead market function for certain electronics products.
4. A push by policy towards the relocation of advanced and high value-added activities in return for the granting of production licenses or market access.

### Empirical Results on Organisational Transitions: The Governance of Supplier Relations and Cross-Border Production

The second, empirical part of this chapter analyses the organisational transition that co-occurred with the spatial transition. Despite the separated presentation of spatial and organisational patterns, the two transitions are highly interdependent in several cases. The cross-cutting issues will be discussed in the concluding section of this chapter. The analysis will focus on governance modes of supplier relations within the region.

Governance of supplier relations is measured by four different modes which are related to the classification used by transaction cost economics. Having one's own production site is the most hierarchical governance mode, while the pure placement of individual orders without any further equity or contractual relation refers to pure spot market relations. Equity joint ventures and contractual (non-equity) framework agreements try to cover the wide range of hybrid arrangements. Firms were asked in the survey to refer to their most important supplier, and the results shown in Table 12.7 are limited to suppliers within the PRD in order to exclude the impact on the results of different institutional environments in other regions in China or abroad. In a previous analysis of the impact of localised institutions on governance modes, Meyer et al. (2009) have already shown that local differences in institutional settings have a major impact on the configuration of inter-firm relations of firms in HK.

**Table 12.7    Governance of relation to the main supplier in the PRD by firms in HK and the PRD**

| | | Type of relation | | | | |
|---|---|---|---|---|---|---|
| Type of partner | Year of firm foundation | own production site | joint venture | framework agreement | only linked by orders | chi-square sign. |
| **All firms in HK** | | | | | | |
| PRD supplier | 2000 and earlier | 68% | 16% | 9% | 7% | 0.000 |
| n=104 | 2001 and later | 40% | 17% | 25% | 19% | |
| **All firms in the PRD** | | | | | | |
| HK customer | 2000 and earlier | 20% | 12% | 8% | 60% | 0.131 |
| n=85 | 2001 and later | 16% | 8% | 4% | 72% | |
| **Chinese-owned firms in the PRD** | | | | | | |
| PRD supplier | 2000 and earlier | 14% | 3% | 14% | 70% | 0.824 |
| n=91 | 2001 and later | 9% | 4% | 17% | 70% | |
| **Foreign-owned firms in the PRD** | | | | | | |
| PRD supplier | 2000 and earlier | 8% | 14% | 24% | 55% | 0.002 |
| n=69 | 2001 and later | 11% | 6% | 6% | 78% | |

*Source*: own survey

The 'front shop, back factory' model applied by firms from HK when they relocated their production facilities to the PRD was strongly based on hierarchical governance modes. Most firms from HK set up their own production sites or entered joint venture agreements with local partners as required by Chinese regulations in the early days of the reform process. This pattern is still visible in the governance modes of HK firms to their PRD supplier in Table 12.7. However, the relation between HK enterprises and domestic enterprises in Guangdong has changed from a Hong Kong-dominated mode to a more cooperative one (FHKI 2007). The survey data reveal that younger firms choose non-equity relations (framework agreements and market-based orders) as often as equity relations, while more than 80 per cent of the older firms still deal with their suppliers via equity relations.

Three major reasons for this transition were mentioned in the interviews with large electronics firms in HK. First, the modularisation of the electronics value chain and shorter product life cycles are requiring firms to organise the sourcing process in a more flexible way. It is therefore a potential competitive advantage to outsource production activities to specialised manufacturing service firms. Second, the capabilities of producers in the PRD have been growing over time. The Li & Fung Research Centre (2006) concluded that the region is offering a rather complete cluster of electronics production capabilities. Especially for low and medium-technology supplies, domestic firms are chosen as suppliers. Third, the institutional change in the region has clearly enabled the reduction of hierarchic modes of governance. It was acknowledged by the interviewees that the legal framework for doing business in the PRD has become more reliable and efficient over time. As predicted by transaction cost economics, high transactional uncertainty, which is the result of deficient institutional constraints, is related to vertical integration within the boundaries of a firm (Williamson 1998). If uncertainty is reduced, spot transactions are more likely to emerge. Simultaneously, asset specificity is reduced in the modular value chain due to a high standardisation of the interface between contract manufacturers and part suppliers (Sturgeon 2002). Less specific assets are another reason for the transition of vertical integration into spot transactions.

Within the PRD, a similar pattern can be observed for supplier relations of foreign-owned firms. They are subsequently reducing equity and contract-based forms of governance while spot transactions are used significantly more often. The underlying reasons are most likely the same as for firms from HK. In addition, several supplier relations of foreign firms in the PRD have been governed by equity and non-equity joint ventures because official regulations in the past required the involvement of local partners when setting up production plants. Therefore, the decreasing importance of joint ventures is also an outcome of more liberal regulations.

Surprisingly, domestic firms in the PRD did not transform the governance of their supplier relations over time, but have always relied on market-based agreements to a large extent. Reasons for this distinct pattern might include the low-tech orientation and thus even lower asset specificity of domestic firms, the absence of regulations that forced foreign firms to form joint ventures with local partners in the past, and the ability to deal with existing institutional uncertainties through other informal safeguards, for example personal relations such as family networks between customers and suppliers. More recently, differences in governance modes between foreign and domestic firms have disappeared, which is an indication that the institutional transition really provides a more level playing field for all players in the PRD.

## Conclusions

It has been the aim of this chapter to describe and explain recent spatial and organisational transitions in the electronics industry of the GPRD. The results have been presented separately for the spatial and the organisational perspectives. After a summary of the major empirical findings, strands of interrelation between the two perspectives will be discussed in this section. Finally, policy recommendations and options for further research will be presented.

*Global-local transition*: As expected from the literature on the localisation of modular value chains in emerging economies, OEM is still a very common mode of production in the GPRD, but a gradual transition towards ODM/OBM activities seems to be taking place. However, many firms sustain their OEM business simultaneously due to its greater stability. Firms in the PRD are moving into the ODM/OBM market more quickly than HK firms and are localising their activities more intensively in the domestic market, while some HK firms are successfully becoming contract manufacturers on a global scale and are now in charge of organising parts of the electronics value chain within Asia.

*Cross-border transition*: Foreign capital, especially from HK and Taiwan, remains of major importance in the PRD and is still responsible for two-thirds of the value added in the electronics industry. But the empirical evidence suggests that the role of HK as a major hub for sales and sourcing in the PRD is declining or at least transformed. Chinese firms are most often embedded domestically and only foreign firms are still using HK for their global sales as often as direct contacts to global customers. Very few companies are locating their future-oriented innovation activities in HK, but prefer a location in mainland China or, in the case of foreign firms, at their home base.

*Transition of governance modes*: The HK-dominated 'front shop, back factory' model, with its hierarchical and complementary cross-border relations, has been transformed into a more competitive partnership over time which is increasingly based on governance via market relations. The major reasons for this transition are the need for flexibility within the global electronics value chain, the improved capabilities of domestic firms and the reduced uncertainty in business operations due to an improved institutional environment. Nevertheless, capabilities and institutions seem to be much less well developed in peripheral locations of the PRD.

*Interrelations of spatial and organisational transitions:* On the one hand, localised institutions in the PRD have shaped the organisational pattern of the electronics industry, which has evolved over time and differs between the two locations studied. The provision of a more complete institutional environment for doing business has resulted in a reduced use of hierarchical governance and private contacts and a rise in the importance of business networks and market-based governance. On the other hand, each distinct mode of organisation results in a specific spatial pattern. Hierarchical governance favours locations in close proximity to HK which enabled a frequent and personal supervision of the production activities on the mainland. Governance at arm's length allows for a relocation of production activities to distant cities within the PRD. The necessity of relying on private contacts pre-defined the locations that were accessible for setting up production facilities. More recently, a further spatial spread of economic activities has been made possible by an increased use of business networks and impersonal modes of contact which allow for locational decisions based to a greater degree on hard economic factors. However, localisation advantages are now in place within developed clusters and might result in a persistence of existing spatial patterns.

*Policy recommendations*: Two major policy recommendations are derived from this chapter. First, policy makers in the PRD are encouraged to substantiate further the institutional transition towards a reliable setting for market-based transactions. The progress in this field already enabled the termination of hierarchical control over production activities in the PRD in the past and has resulted in more independent development and capability-building of PRD producers. Second, policy makers in HK are advised to take deliberate action to refocus and intensify those areas in which they still possess a competitive advantage over the PRD. Considering the fact that sales and sourcing is increasingly carried out by firms in the PRD, and that innovation activities were never located in HK to a significant degree, the city should further concentrate on advanced business services and its role as an organiser of global value chains in order to remain an important complementary location for firms situated in the PRD. Alternatively, HK might take further measures to position itself as a global city in its own right without a particularly strong connection to its hinterland.

*Necessity for further research*: While this chapter has presented empirical findings from two cities which are among the most developed in the PRD, further research may be able to trace recently emerging developments in peripheral locations of the PRD and check whether the organisational transition of previously industrialised locations is reflected there due to a lack of institutions, or whether the overall improvement of capabilities results in a spread of networked and market relations from the very beginning. A combination of large-scale investment data with survey data could be useful to further substantiate the findings on the spatial transition of the HK-PRD link. The focus on governance modes and its integration with clustering and upgrading has proven its potential to be applied in further studies which could take institutional factors into consideration even more explicitly to explain the observed transitions.

## Acknowledgements

The research underlying this chapter has received funding from the German Research Foundation (DFG) within the framework programme SPP 1233 'Megacities – Megachallenge: The Informal Dynamics of Global Change'.

## References

Barney, J.B. and Zhang, S. 2009. The Future of Chinese Management Research: A Theory of Chinese Management versus A Chinese Theory of Management. *Management and Organization Review*, 5(1), 15–28.

Bellandi, M. and di Tommaso, M. 2005. The Case of Specialized Towns in Guangdong, China. *European Planning Studies*, 13(5), 707–28.

Enright, M., Scott, E. and Chang, K.M. 2005. *Regional Powerhouse: The Greater Pearl River Delta and the Rise of China*. Singapore: Wiley.

Ernst, D. 2005. Pathways to innovation in Asia's leading electronics-exporting countries – a framework for exploring drivers and policy implications. *International Journal of Technology Management*, 29, 6–20.

FHKI 2007. *Made in PRD – Challenges & Opportunities for HK Industry*. Hong Kong: Federation of Hong Kong Industries.

Guerrieri, P. and Pietrobelli, C. 2004. Industrial districts' evolution and technological regimes: Italy and Taiwan. *Technovation*, 24, 899–914.

HKCPU 2010. *Hong Kong's Role in the Development of the Mainland*. Commission on Strategic Development Paper CSD/1/2010. Hong Kong: Central Policy Unit.

HKPC 2003. *From OEM to ODM*. Final Report. Hong Kong Productivity Council.

HKTDC 2008. *Study on OEM, ODM and OBM: Extending the Supply Chain with Added Value*. Hong Kong Trade Development Council.

Hobday, M. 1995. East Asian Latecomer Firms: Learning the Technology of Electronics. *World Development*, 23(7), 1171–93.

Hürtgen, S., Lüthje, B., Schumm, W. and Sproll, M. 2009. *Von Silicon Valley nach Shenzhen. Globale Produktion und Arbeit in der IT-Industry*. Hamburg: VSA-Verlag.

Langlois, R.N. 2003. The vanishing hand: the changing dynamics of industrial capitalism. *Industrial and Corporate Change*, 12(2), 351–85.

Li & Fung Research Centre 2006. *Industrial Clusters in Pearl River Delta (PRD)*. Industrial Cluster Series 2. Hong Kong: Li & Fung Group.

Li, Y. and Peng, M.W. 2008. Developing theory from strategic management research in China. *Asia Pacific Journal of Management*, 25, 563–72.

Lu, L. and Wei, Y.H.D 2007. Domesticating Globalisation, New Economic Spaces and Regional Polarisation in Guangdong Province, China. *Tijdschrift voor Economische en Sociale Geografie,* 98(2), 225–44.

Luo, J. and Wei, Y.H.D. 2009. Modeling Spatial Variations of Urban Growth Patterns in Chinese Cities: The Case of Nanjing. *Landscape and Urban Planning,* 91(2), 51–64.

Markusen, A. 1996. Sticky Places in Slippery Space: A Typology of Industrial Districts. *Economic Geography,* 72(3), 293–313.

Markusen A. 1999. Fuzzy concepts, scanty evidence and policy distance: The case for rigour and policy relevance in critical regional studies. *Regional Studies,* 33, 869–86.

Mathews, J.A. and Cho, D.S. 2000. *Tiger Technology: The creation of a semiconductor industry in East Asia.* Cambridge: Cambridge University Press.

Meyer, S., Schiller, D. and Revilla Diez, J. 2009. The Janus-faced economy: Hong Kong Firms as intermediaries between global customers and local producers in the electronics industry. *Tijdschrift voor Economische en Sociale Geografie,* 100(2), 224–35.

Moulaert, F. and Sekia, F. 2003. Territorial innovation models: A critical survey. *Regional Studies,* 37, 289–302.

Park, S.O. 1996. Networks and embeddedness in the dynamic types of new industrial districts. *Progress in Human Geography,* 20(4), 476–93.

Park, S.O. 2000. Regional Issues in the Pacific Rim. *Papers in Regional Science,* 79, 333–42.

Poppo, L. and Zenger, T. 2002. Do formal contracts and relational governance function as substitutes or complements? *Strategic Management Journal,* 23(8), 707–25.

Powell, W.W. 1990. Neither Market Nor Hierarchy: Network Forms of Organization. *Research in Organizational Behavior,* 12, 295–336.

Sturgeon, T.J. 2002. Modular production networks: a new American model of industrial organization. *Industrial and Corporate Change,* 11(3), 451–96.

Sturgeon, T.J. and Lester, R.K. 2003. *The new global supply-base: new challenges for local suppliers in East Asia.* IPC Working Paper Series 03-006. Cambridge: MIT.

Wang, J. and Mei, L. 2009. *Dynamics of labour-intensive clusters in China: Relying on low labour costs or cultivating innovation?* Discussion Paper 19509. Geneva: International Institute for Labour Studies.

Wang, Y. and Nicholas, S. 2007. The formation and evolution of non-equity strategic alliances in China. *Asia Pacific Journal of Management,* 24, 131–50.

Wei, Y.H.D., Li, W.M. and Wang, C.B. 2007. Restructuring Industrial Districts, Scaling Up Regional Development: A Study of the Wenzhou Model, China. *Economic Geography,* 83(4), 421–44.

Wei, Y.H.D., Lu, Y.Q. and Chen, W. 2009. Globalizing Regional Development in Sunan, China: Does Suzhou Industrial Park Fit a Neo-Marshallian District Model? *Regional Studies*, 43(3), 409–27.

Williamson, O.E. 1998. Transaction cost economics: How it works; where it is headed. *De Economist,* 146(1), 23–58.

Yang, C. 2007. Divergent Hybrid Capitalisms in China: Hong Kong and Taiwanese Electronics Clusters in Dongguan. *Economic Geography,* 83(4), 395–420.

Yang, C. 2009. Strategic Coupling of Regional Development in Global Production Networks: Redistribution of Taiwanese Personal Computer Investment from the Pearl River Delta to the Yangtze River Delta. *Regional Studies*, 43(3), 385–407.

Yeung, H.W.-C. (ed.) 2006. *Handbook Of Research On Asian Business.* Cheltenham: Edward Elgar.

Yeung, H.W.-C. 2009. Regional Development and the Competitive Dynamics of Global Production Networks: An East Asian Perspective. *Regional Studies,* 43(3), 325–51.

Yeung, H.W.-C. and G.C.S. Lin 2003. Theorizing Economic Geographies of Asia. *Economic Geography,* 79(2), 107–28.

Zenger, T.R., Lazzarini, S.G. and Poppo, L. 2002. Informal and formal organization in new institutional economics. *New Institutionalism in Strategic Management,* 19, 277–305.

Chapter 13

# Economic Competitiveness, 'Glurbanisation' and 'Soft' Spatial Interventions: Insights from Four Australasian Cities

Steffen Wetzstein

## Introduction

How do new local-global economic patterns correspond to new forms of territorial economic management? How do actors and places at the 'edge' of the world attempt to reposition themselves favourably in a competitive globalising world? This chapter addresses these questions by problematising the nexus between economic competitiveness, urban-centred globalization and territorial governance strategies in the context of Australasia (a term standing for New Zealand and Australia). It draws on the 'industrial transition' proposition of this book to ascertain to what degree changes in the nature of contemporary investment, economic and labour processes are mirrored by altered strategies of cities and regions to influence the investment environment. Mindful that the proposed 'industrial transition' thesis (fuzzy and volatile international division of labour, local-global economic interactions and coordination, general – but not complete – loss of western economic leadership) stems mainly from restructuring experiences in the early industrialised countries, this study thus tentatively tests the applicability of this framework for understanding economic and political change in later developed, non-core western countries.

This inquiry is conceptually premised on the assumption that the pursuit of global competiveness for economic and territorial actors has become the dominant governance ethos in the early 21st century (Bristow 2005). It also recognises that economic and urban interventions in the name of 'competitiveness' are intimately tied to the issue of how private investors can actually be governed in their decision-making and resource allocation under market-facilitative, neoliberal conditions (the global financial crisis and the EURO-crises gave painful evidence to this dilemma). Changes in the nature and direction of territorial governance are explored in the context of Australia and New Zealand (Australasia). This region can easily be overlooked in largely Northern Hemisphere-centred economic geography and critical urban/regional studies because of its small population size and its semi-peripheral location on the world map. Yet, Australasia clearly emerged out of the global financial crisis as one of the few economic winners. Both, its strong resource-based export sectors (e.g. mining, oil/gas, agriculture, horticulture) and emerging post-industrial, urban-centred industries (e.g. finance, business services, higher education, knowledge-industries) are increasingly well connected with global markets in general, and emerging Asian markets, in particular. It can be expected that territorial intervention strategies in this part of the world show interesting similarities but perhaps also marked differences in comparison with the large industrialised regions of North-America, Europe and East Asia.

Territorial economic governance can be approached from two complimentary political economy perspectives. First, it is said to incorporate not just state actors but also business and community actors, has moved from nation-state arrangements to sub-national and supra-national scales, is linked in complex international policy regimes and is often distant and removed from the spheres where actual investment decisions are made (Swyngedouw 1997, Brenner 2004). This strand of academic work understands governance as an arena of 'hard' interventions; asking questions about who is influencing whom in policy and political settings, which actor-configurations can be identified, which dominant scales of operation and power be discerned, and which specific forms of negotiation and reconciliation be observed. A second literature strand deals with the 'soft' aspects of interventions, examining the often mundane and largely discursive governing practices, particular political rationalities and specific governmental techniques that underlie attempts to govern, influence and manage 'at-a-distance' (Larner, Le Heron and Lewis 2007). The key concern here is with the 'how' of interventions; emphasising coordination through specific technologies of government such as partnerships, the role of informal and fluid political practices and the use of narratives, benchmarking and associative practices. Importantly, this governmentality-informed thinking is particularly sensitive to the role of context and conditions in the constitution of new governing arrangements. The analytical benefits of this 'soft' approach make it highly suitable to investigating 'the making' of local – global territorial relations involving actors and places in Australasia.

The methodological strategy of this chapter is based on the assumption that cities are particularly important sites in the age of globalization. This significance of the urban scale has been recognised in Jessop and Sum's (2000) work on 'glurbanisation'. In contrast to global firms' 'glocalisation' strategies that seek to exploit local differences to enhance their global operations, the 'glurbanisation' label refers to 'entrepreneurial strategies that are concerned to secure the most advantageous insertion of a given city into the changing interscalar division of labour in the world economy' (Jessop and Sum 2000: 2295).

Contemporary 'glurbanisation' strategies are comparatively examined for Australia's (population: 22.7 million) and New Zealand's (population: 4.4 million) highly urbanised societies, in which big cities predominantly serve as commercial hubs, service centres and gateway locations rather than as sites of manufacturing goods. Both countries experienced far-reaching neoliberal restructuring in the 1980s and 1990s that introduced market-friendly policies, privatisation and opening of borders to global investment and people (Le Heron and Pawson

1996). The research emphasis is on four cities across these two countries; Perth, Melbourne, Sydney and Auckland (see Figure 13.1).

Perth (population: 1.7 Mill) is the most isolated capital city in the world. Its economic base is overly reliant on the resources sector, meaning that investments, business confidence and population growth are strongly linked to the boom and bust patterns of this sector (Kennewell and Shaw 2008). Melbourne (population: 4.0 Mill) is said to have been turned around from a declining manufacturing centre in the late 1980s to a self-proclaimed global service, knowledge and event centre 20 years later (O'Connor 2003). Sydney (population: 4.5 Mill) is Australia's largest economic and commercial centre, and widely perceived as its prime global city. This perception is largely based on the number of headquarters of transnational and mostly financial companies, the sheer size of the economy, and the global imagery fuelled by world-class local events (McGuirk 2007). Auckland (population: 1.4 Mill), finally, is New Zealand's economic centre and traditional gateway to the world. Today, it is perceived as New Zealand's only global city and therefore deserving particular political and policy attention (Wetzstein 2008a).

The empirical emphasis is on the variety of 'soft' spatial governmental strategies that can be observed across these four sites, and how they help to constitute globalising economic intervention fields. These interventions include circulating spatial imaginaries, differing discursive attempts to incorporate actors and places into global resource flows, and the underlying practices that make spatially expanding investment and economic processes seem thinkable and realisable. In other words, the research goal is to investigate how 'glurbanising' Australasian cities are imagined and made via, at least initially, discursive forms of spatial influence.

This chapter is organised around two linked questions that speak to the overall concern of distinguishing a region-specific governance direction that facilitates global economic competitiveness. First, it asks what explicit spatial governmental strategies can be discerned in attempts to 'glurbanise' Australasia, and, second, how are particular city-specific conditions and context affecting such strategies? In critical political economy tradition, the answers are sensitive to emerging tensions, observable mismatches and problematic trade-offs across space and time. The choice of the four case study sites has been influenced by the author's particular positionality of being involved in academic and policy knowledge production in Auckland and Perth. Sydney and Melbourne have been chosen as investigative sites because of their alleged importance as Australia's economic powerhouses, centres of political power and key nodes of globalising processes. Using an ethnography-based methodology[1], it is argued that 'glurbanisation', rather than other forms of

---

1    This study is based on extensive policy document review, broad media scanning, a range of personal work experiences, observations and conversations at the policy-academic interface as well as comprehensive in-depth interviews with 44 senior governance actors from public and private sectors in the four cities. Interviewed policy makers work largely in planning, strategy, economic development and research roles, while business leaders,

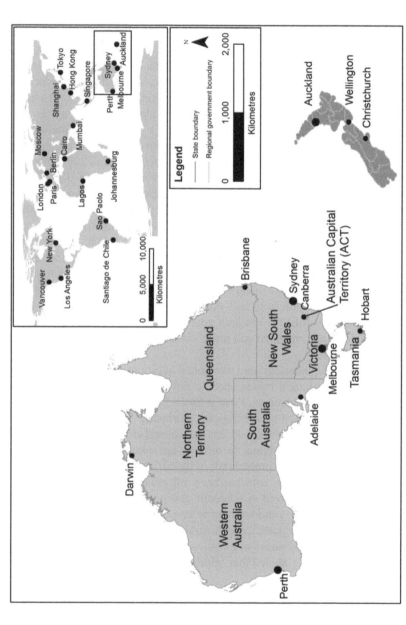

**Figure 13.1  Australasian case study sites**

economic repositioning such as technological upgrading or investments in green industries, is central to enhancing the region's global competitive standing.

In detail, this chapter claims, first, that 'glurbanisation' involves a range of initially discursive strategies that aim to bring the 'global' and 'local' in imagined proximity in order to constitute potentially powerful economic local – global relationships and interactions, and second, that cultural-institutional context and socio-economic conditions in each place mediate the extent and contours of these interventions. A discussion of key strands of the international literature is followed by five short empirical sections that engage with particular spatial governmental strategies that (potentially) link Australasian places, spaces, actors and resources in the circuits of globalising capitalism. Concluding thoughts call for more critical engagement with the uneven effects of global repositioning attempts.

## Economic Competiveness, Urban Entrepreneurial Governance and Globalising Spatial Strategies

Territorial economic governance under neoliberalising and globalising conditions has been explored from a variety of angles in the international literature. One influential strand stresses the up-scaling and down-scaling of previously nation-centred governance (Swyndgedouw 1997), while other academic work emphasises the broadening of actor-coalitions beyond the state (Jessop 2002). Increasingly, attention has been paid to the need for actors and places to enhance their competitiveness vis-a-vis other places via new forms of innovative strategies. So is it now common for cities, regions, and nations to assess, improve, and publicise their competitive standing. Indeed, competitiveness has become a key marker for political strategies and a policy buzzword concerned with politically managing national and regional economic transformation in many countries. Cities and city-regions are said to become particularly important objects of economic governance because they are key nodes of competitive activity, centres of power and exchange, sites of entrepreneurial governance, attractive to global capital in production and consumption circuits, the most appropriate planning governance level, laboratories for developing sustainable practices as well as producers, distributers and consumers of mobile and modifiable policy models. They are also viewed as nodes through which particular ideas about the world circulate, places where policies are formulated, contested and implemented, and sites where particular outcome patterns are envisioned and pursued. In addition, questions of economic prosperity, liveability, sustainability and citizenship are often linked to urban places.

---

corporate managers and professionals are from mining, higher education and accountancy sectors as well as various business associations, urban development organisations and urban think-tanks. For more methodological detail, see Wetzstein (2010).

'Glocalisation' has been described as a key state spatial strategy to facilitate capital accumulation and economic growth in the era of globalising competitive capitalism (Benner 2003). Jessop and Sum's (2000) notion of 'glurbanisation' can be construed as a particular expression of such re-scaling processes. They utilise this term to theorise the 'Siliconisation' strategies, or the imagining of, and investments in, information-technology futures, of Asian cities such as Hong Kong and Singapore as points of global difference. Hodson and Marvin (2007) claim that understandings of the differential positionings and repositionings of territories at particular urban and metropolitan scales within multi-scalar networks (Brenner 2004) requires a sensitivity to agency in the ways in which particular places engage with these growth agendas. Matusitz (2010) contends that 'glurbanisation' has now achieved the status of an exemplary framework in urban development strategies. Importantly, this increasingly taken-for-granted policy approach works with the assumption that there is a competitive global urban system. This imagined global hierarchy of cities involves an awareness of the economies, wider living arrangements and policy approaches of other cities especially via benchmarking practices. In this context, Sassen (2009) claims that the diverse specialised economic capabilities needed in a global world are produced in the complex and thick environments of particular 'global cities'; places that function as bridge between their national economies and the global economy. Yet interestingly, global firms need groups of cities, meaning that cities do *not* simply compete with each other. In sum, cities are central to globalising investment and economic processes, and therefore mediating institutions, policy settings and governing strategies involving these places need to be investigated in more detail.

Scholarly attention has also shifted to the intriguing question of *how* governing is achieved under conditions of market-facilitative regulation, a financially constrained public sector and rising local-global interdependencies. This concern with the constitutive dimensions of governance, or how interventions come into existence, expand, and evolve, has been addressed by a post-structural political economy perspective (PSPE) (see Larner, Le Heron and Lewis 2007). Stemming from the analytical engagement with emergent post-restructuring[2] governance structures in New Zealand, it highlights governance and economic processes as contingently 'in the making' (Wetzstein and Le Heron 2010), featuring emergent alignments, regularities and arrangements, and having constitutive effects. By paying attention to both discursive and material governance dimensions, PSPE has shown particular interest in the possibilities and complex ways of governing a multitude of diverse actors at-a-distance.

In line with a renewed interest in the nature of practice in economic geography, emerging *discursive* practices of governance have received particular attention in this regard (Wetzstein 2008b). These strategies refer not only to all the ways

---

2    Economic restructuring refers to the wide-reaching effects of the well-publicised comprehensive neoliberal political reforms in New Zealand in the 1980s and 1990s. For a detailed account, see Le Heron and Pawson (1996).

in which actors communicate with each other but to complex networks of signs, symbols, and practices through which the world becomes meaningful to people. They encompass combinations of a range of practices, including 'story telling' to inspire, motivate and mobilise other actors, associative practices such as networking and project management to forge new productive linkages between people and resources, the use of benchmarking to create globalising imaginaries for local actors and the proliferation of indicators to constitute 'self-reflexive' actors. The discursive framing of governance issues and solutions is important as, drawing on certain aspects of materiality and experience of everyday life, these governmental techniques allow attention of a wide range of people to be focussed on a common concern so as to achieve a particular political purpose.

A diverse set of literatures has illuminated the roles of various governmental techniques in reconstituting territorial relationships and interactions across space. The concept of *imaginaries* has been used to describe how academic and policy ideas become strategic intellectual projects that sometimes go into extended circulation. Recent work focussed on their mobilisation, travel, translation, apprehension and contestation (Wetzstein and Le Heron 2010). In her work on the normalisation of Toronto as a 'city-regional' space, Boudreau (2007) refers to the discursive power of spatial imaginaries as people's mental maps and collectively shared internal worlds of thoughts and beliefs that structure everyday life. Calculative practices such as benchmarking, audits, evaluations, indicatorisation as well as league tables and ranking exercises are viewed as particularly powerful in the making of new conceptions of global spaces and subjects by bringing together previously heterogeneous and spatially disparate economic objects. Put differently, they are central to the growing toolset of *global referencing capabilities* that involve scanning knowledges and practices across space (Larner and Le Heron 2005). McCann (2008) considers spatial imaginaries, expertise and numerous apparently mundane practices such as calculation and communication in order to examine *urban policy mobilities* of people, products, and knowledge in global circulation. The enhancement and promotion of *urban images* and the establishment of unique city brands also feature prominently in governance strategies of western cities (Kavaratzis 2005). However, solid evidence for the link between how people perceive a certain place, and resulting decision-making that incorporates or excludes this location, is often lacking. Finally, Larner, Le Heron and Lewis (2007) claim that in tandem with international competitiveness becoming New Zealand's key object of economic governance in recent times, intensifying *global connectedness* has become the country's prevailing governmental rationality. In practice, this move entails the framing of economic and social policy around increasing local-global connections and the constant drive to generate economic and cultural values for the global market place.

## Spatial Imaginaries: Constituting Local-Global Territorial Relations

A first spatial governmental strategy directed at raising territorial economic competitiveness entails the purposeful creation of new, or altering existing, spatial imaginaries. These 'mental maps' of actors are now increasingly oriented towards a global frame of reference. Discursive constructions such as Perth as Australia's Face to Southern Asia, Auckland's identity as Asia-Pacific Rim city or Sydney's location in the Asia-Pacific Region illuminate attempts to reposition actors and places in regards to the 'global'. These imaginaries highlight that the 'global' for a competitive Australasia has much to do with building links and relationships with Asian actors and places. New spatial imaginaries are also created by the (re)packaging of places as, for example, global AND green (Sydney) and global AND liveable (Melbourne). In these examples urban-based corporate interests and policy makers actively construct combinations of desirable spatial and societal representations that facilitate understandings of competitive urban economic processes as reconcilable with social and environmental objectives. Thought to be well-suited to attract skilled overseas labour and quality capital, they also provide the justification for particular local policy and spending decisions. But these discursive representations are necessarily one-sided, over-emphasising desirable futures and masking inevitable tensions and contradictions in the process. These inconsistencies are, for example, expressed in mismatches in governmental attention and subsequent public and private resource allocation between Australia's and New Zealand's cities on the one hand, and the surrounding rural and semi-rural parts of these countries on the other hand. In some respect, this local spatial unevenness seems to be the other side of the globalization coin. Unsurprisingly, actors representing larger spatial jurisdictions (e.g. Australian States) seek reconnections across urban-rural spaces. The 'One Victoria' initiative by the Victorian Competition and Efficiency Commission and the 'Perth Vibrancy and Regional Liveability' study by the Western Australian Chamber of Commerce are just two examples that underscore the desire to respond to the uneven effects of intensifying local-global economic flows on people and places.

An interesting example of how spatial imaginaries are re-framed for global repositioning is the concept of *time zone*. In the recent past, and in particular after decreasing economic protection in the wake of the neoliberal reforms, access to new export markets has become a more significant policy ambition. In this context, Australia's and New Zealand's location at the 'bottom' of the world and distant to large Northern Hemisphere markets was viewed as disadvantage. In today's globalising context, however, location framed relationally in form of 'time zones' by some Australian elite actors means re-packaging it as economic opportunity rather an investment barrier. To these interests, Australia is well-positioned to serve North-American and European clients by taking advantage of the time difference that allows work to be done here while people there sleep or spend leisure time. Being the first major financial market to open each day, Sydney in particular is said to have a significant time zone advantage over other Asia-Pacific locations. It also provides a time zone bridge, linking the closing of the US and the opening of the European markets.

The so-called 'IN-THE-ZONE' conference in Perth, a major international public policy conference initiated by the University of Western Australia in conjunction with the Western Australian State Government, large mining businesses and a wide range of business organisations and the media in Spring 2009, featured 'time zone' as a catch-phrase to stimulate debate on the public policy, geo-political and business challenges facing Australia and its major trading partners in China, Japan, India, South-Korea and South-East Asia. Using the catchy phrase 'THE ZONE', this high-profile conference tried to boost perceptions of the region's favourable location in respect to Asian markets. Indeed, Perth was constructed as 'a leading city IN THE ZONE [which it shares] with 60 per cent of the world's population and the nations promising the greatest economic growth of the twenty-first century' (EECW 2009).

It can be argued that by strategically reconsidering the location within a certain time zone, space and time are framed simultaneously (McCann 2003) as relational economic resources for political ends. Jessop and Sum's 'glurbanisation' concept explicitly recognises how time as well as space is governed in the production of urban-based competitive advantage in a globalising world. Referring to chronotopic governance, they argue that time and space are re-articulated for structural or systemic competitive advantages. The re-conceptualisation of spatial isolation as territorially framed problem to relationally framed solution in order to access global resource flows is a recent example of how this chronotopic governance is constituted in daily practice. It is important to note that not all case study sites are profiting evenly from the circulation of new spatial imaginaries. Auckland, for example, gained global policy recognition for its radical political amalgamation under the 'Super-City' label, but 'time zone' strategies proved rather unsuccessful there for reasons of spatially fragmented exports markets and missing critical mass in potentially globalising industry sectors.

### Calculative Referencing Practices: Benchmarking and Ranking

A second spatial governmental strategy thought to facilitate global economic competiveness is the rising use of calculative spatial referencing practices such as benchmarking and the use of league tables. These number-based exercises clearly are becoming more central to the constitution of new spatial objects of territorial economic governance. One example is a recently published 'Global City' benchmarking exercise commissioned by the 'Committee for Sydney'. This report selected particular cities as benchmarking categories in regards to particular dimensions of global urban competitiveness. So define London, Tokyo and New York Sydney's global aspirations and standing, Singapore, Hong Kong and Shanghai are seen as direct competitors, and Vancouver, San Francisco and Los Angeles are depicted as places that have similar urban characteristics and challenges to Sydney (Committee for Sydney 2009). Interestingly, European cities where many Sydneysiders of this and past generations originated from do not feature at all. Sydney's globally connected corporate business community thus shapes particular selective conceptions about the city's local-global relationships, about which places and actors should serve as reference points, and who should be excluded. The intervention-space constituting character of calculative referencing practices is also revealed by the geography-centric nature of some of the chosen indicators in this study such as 'percentage of resident population born overseas', 'air transport passengers' or 'top global fortune 500 companies in the city'. Some benchmarking indicators are benchmarks themselves, for example, the Shanghai Jiaotong university ranking (number of top 500 universities), the Anholt City Brand ranking and the Economist Intelligence Unit's (EIU) liveability survey index score.

This 'benchmarking of benchmarking' approach raises questions about the credibility of this practice as much contextual and empirical detail is under threat of getting lost in these multiple discursive translations. It is telling that benchmarking has been more dominant in Sydney's policy world compared to the three other case study sites; an indication perhaps of the more globalised and corporatist private sector in this city that is able to put this originally commercial management tool to best political use. The author found that diverse benchmarking reports, own policy work experiences and interview messages are remarkably aligned in two respects. Number- and indicator based techniques are becoming more common for spatial referencing in Australasia, and the current North-American and East-Asian spatial emphasis in comparative policy work actively constitute the Asia-Pacific-Region and Asia-Pacific-Rim as the increasingly normalised frame of reference for competitive urban and economic processes across the region.

Quality of Life surveys and so-called liveability studies are the other form of ranking exercise that seems to be particularly powerful in the constitution of new governance discourses, and indeed deeper entrenched spatial imaginaries. They are highly prone to attention from the media and politicians, especially in Australasian cities that are often highly rated in annually published international liveability studies. These rankings have not only elevated 'liveability' to a fashionable policy buzzword, but indeed have resulted in a substantial trail of policy action, political initiatives and technical reports in some cities. Its success in Australasian urban policy circles can nowhere be shown more clearly than in Melbourne's policy world in the mid-to-late 2000s. On the back of favourable liveability ratings, policy makers on local and state levels actively re-interpreted and deliberated what this term may mean for Victoria and its capital city-region. Institutions such as the 'Committee for Melbourne', the City of Melbourne, the Victorian Competition and Efficiency Commission, the Melbourne Growth Areas Authority, the Future Melbourne project and the Coalition for People's Transport engaged in a dynamic policy debate that stretched understandings of liveability and demarcated its definitional boundaries.

A much less enthusiastic political and policy response can be expected from national ranking exercises that put the spotlight on urban management problems such as the independent assessment of planning systems of Australian Capital Cities and the newly created Sustainable Cities Index by the Australian Conservation Foundation. It needs to be stressed that political and economic decision-makers as well as academic, policy and media commentators strategically use number-based techniques to further their claims about the world, where resources should be spend, and towards which ends. In this sense, numbers are far from neutral but highly political. The increasing emphasis on number-based comparative techniques in economic and urban policy and politics can be critiqued for its rarely questioned methodologies, seldom explicitly acknowledged details about content and lack of any in-depth analyses of index-weighing formula. But it is arguably the actual governmental 'work' of calculative referencing techniques in people's minds that must be examined in more critical detail by the research community.

## Global Resources and Model Search: Scanning, Attracting, Importing

Global best practice scanning, resource attraction and urban policy import can be construed as a third pathway to building economic competitiveness on the basis of local–global referencing. Interviewees' answers in all four case study sites highlight the importance of actors looking to places in the 'global' for guidance on 'successful' development. These elite actors 'pick and choose' places that are perceived as doing well in particular policy areas and elevate those to role models worth emulating. In the land-use policy area, for example, ideas on compact city design, mixed-use developments and alternative transport modes have strongly informed planning policies in Australasian cities.

These ideas are largely sourced from Northern Hemisphere cities and their policy experts. One of those is Copenhagen's star architect Jan Gehl who has been visiting the region for a long time and influenced, to varying degrees, inner city and waterfront redevelopments in places such as Perth, Adelaide, Melbourne, Sydney, Hobart and Christchurch. In contrast, Transit-Oriented Design (TOD) and 'new urbanism' design principles that affected urban redeveloped projects in many cities. Interestingly, the TOD-train station design in Perth's suburb Subiaco then became the exact blueprint for a large scale intensification project in West-Auckland.[3] Global 'best-practice scanning' also entails the search for 'fit-for-purpose' urban political governance. So is Auckland's new local government structure the result of adapting and assembling ingredients from policy models in Vancouver, London and Brisbane; resulting in a novel combination that has made Auckland's new 'Super City' arrangement in its entirety a unique creation.[4] Yet, the world of policy transfer also incorporates the diffusion of worst-practice such as city images of un-governability, under-investment and in-affordability; policy

---

3    Research interview with the Waitakere City Mayor on 14 April 2010.
4    Research interview with the Chair of the Royal Commission on Auckland Governance on 13 April 2010.

trajectories that are increasingly associated with Sydney. These negative attributes can be used in governance arenas to critique and 'other' undesirable policy options, unsuitable city representations and unwanted institutional arrangements.

Best practice travels through many channels. These include policy makers' in-house desktop research and fact-finding trips, global management consultancies and their internal knowledge transfer, the celebrity circuits of academic gurus as well as the referencing work of intermediary governance organisations such as think-tanks (e.g. the 'Committee for Perth') and powerful global governance entities (e.g. the World Bank). 'Travel agents' also include temporary and permanent skilled labour migration, the personal experiences of elite actors such as politicians, bureaucrats and business people and the influential narratives and images circulating through the popular media. The conducted interviews also point to the particular role of staff members in global policy transfer. So are government personnel and members of the private sector workforce quite often born overseas or educated at prestigious foreign universities.[5]

The associated knowledge transfer can be remarkably effective as in the case of the City of Sydney's chance recruitment of one of the key thinkers behind the Greater London Authority's environmental policies; an appointment that added significantly to the success of the currently implemented 'green city' strategies in Inner Sydney. In sum, it appears that these context-specific urban policy mobilities (McCann 2008) from around the world are becoming important discursive resources for reimagining and changing economic governance in Australasian cities.

The nexus between spatial imaginaries, policy import and seemingly mundane, every day practices can be shown in efforts to promote the city-region as a desirable object of governance for Perth. Here, the University of Western Australia and the 'Committee for Perth' commissioned a benchmarking policy report on metropolitan and local governance in different jurisdictions. This study was then repeatedly revised and repackaged by a public relations company before a press release made it accessible to the media. This publication coincided with a Perth visit of the Chair of the Royal Commission on Auckland's Governance and the CEO of the 'Committee for Auckland'. Meeting key policy makers and business people in Western Australia's capital city, they shared their lessons on the identification of, and transition to, a centralised local governance model for New Zealand's economic centre. Both initiatives in tandem mobilised 'city-regional' thinking in the local government reform debate in Perth that was even acknowledged by the State Minister for Local Government. Interestingly, Auckland is now considered as a new role model for urban reform in the region and challenges to some degree the dominant position of Melbourne as a policy model exporter (Wetzstein 2010). From a theoretical perspective, this example of urban policy import illustrates Wetzstein and Le Heron's (2010) claim that time- and place-specific institutions and their practices critically matter in the mediation of globalising governance arrangements.

---

5    Research interview with a Sydney Policy Maker on 9 March 2010.

## Broadcasting 'Potentialities': Place-Making and Global City-Marketing

The goal of attracting globally mobile resources to Australasian cities can be linked to proliferating place-making, city-marketing and city branding efforts. This fourth set of globalising governmental practices can be understood as advertising the potential value of places' work in the global division of labour. Put simply, it is about broadcasting 'potentialities', or the multiple possibilities of adding value to globalising circuits of capital and labour, and raising visibility on a global scale. This strategy entails high-profile waterfront-redevelopment initiatives, hosting globally noticed events and spectacles, fashioning iconic architecture and high-quality urban design as well as creating new signature institutional arrangements. It also includes new rounds of targeted city-marketing that use discursive resources for the creation of unified city visions or sharp and distinctive city brands.

One can understand this governmental strategy as the projection of recognisable aspects of the 'local' to particular actor groups of the 'global' in order to engender or enhance profitable local-global interactions. An alternative explanation simply views it as powerful justification for advocating changes in local public and private investment. The particular mix of place-making and city-marketing projects differs for each case study site; underscoring the fact that governmental practice strongly depends on local conditions and context. So has Perth struggled for years to find agreement on its CBD-waterfront redevelopment, while Sydney and Melbourne have already reaped the benefits of implementing large-scale waterfront projects. In Sydney, 'branding' has now become a new buzzword since the 'Brand Sydney' project was launched in 2008 in order to develop a global brand logo for the city. While the Greater Sydney Partnership endeavours to 'market Sydney and its existing assets more effectively by shifting current perceptions and stimulating investment, promoting world-class education, increasing visitation (both domestically and internationally) and increasing tourism and trade' (Lee 2010), Melbourne has gained much traction already in many of these activity areas over the last decade.

Auckland and Perth, in contrast, find it hard to develop unifying marketing strategies or even an internationally recognisable brand. In both places efforts are directed to vision-building at the moment; an exercise linked to inspiring the local population and setting the direction for future planning. But while Auckland lacks critical mass in recognisable economic and cultural attributes, Perth's challenge is to shake off its largely negatively interpreted mining 'tag'. Viewed critically, the packaging, marketing and selling of cities as commercial products and associated attempts to press rich social, economic and cultural complexity into tight singular propositions such as Auckland's 'One Plan' – all in the name of raising global competitiveness – must be of concern because it always involves the privileging of particular material interests and their urban representations at the exclusion of other interests and their visions.

Place-selling goes hand in hand with the wide-spread voicing of heightened aspirations and ambitions in almost every policy and institutional domain in

Australia and New Zealand. Increasingly, these desires are phrased in direct regard to the 'global'. This is the geographical scale where broader objectives can be best articulated, specific strategies formulated and attention, expenditures and possible trade-offs' justified. The term 'world-class' is used frequently, and seems to be more commonly used in Australasia than, for example, in the European context. So is Perth currently re-imagined as a 'World Centre for Indigenous Culture and Cultural Experience' and as a place with 'world-class' amenity (Committee for Perth 2010).

According to many promotional websites, Australasian cities boost world-class infrastructure and world-class research and development facilities. Global ambition, it seems, serves as a key reference point, and source of legitimisation, for local institutional and investment changes. To the critical observer, however, claims of 'world-class'-ness are too often not aligned with reality. Just like other phrases that exaggerate in order produce unreflective affiliation and even hype such as 'fantastic', 'great', 'vibrant', 'exciting' and 'unique', they are often the specific discursive dimensions of national/state/city-boosterism aimed at persuading people to buy-in, belief-in and consume certain products, services, packages or perceived advantages. Over recent years, these boosterist discursive techniques have been transferred from its original commercial applications to the now glossy pages of many public policy documents. More generally, and confirming Kavaratzis' (2005) perspective on city-marketing and branding, it seems that the business of shifting perceptions through urban renewal projects, sophisticated marketing campaigns, targeted institutional mechanisms and the constant reiterating of globally-referenced aspirations has become an important part of projecting the Australasian 'local' into the 'global'. In this regard, the broadcasting of urban 'potentialities' is seemingly becoming a normalised spatial governmental strategy for this region.

**Enhancing Global Connectedness: Strategic Network Building**

A fifth and final spatial governmental strategy is strategic network building across space. At local and regional scales, these networks are now more likely than in the past to cross traditional institutional boundaries such as the public-private divide. Multi-actor alliances such as the 'Committee for [City]' entities and the Greater Partnership for Sydney illustrate the power of public-private networking in order to be able to speak in behalf of places, and thus further particular political and policy agendas. Spatial governmental networks are also forming increasingly on national scales (e.g. the Council of Australian Governments), in trans-national settings (e.g. the Australian- New Zealand Local Government Chief Financial Officers Group), and across global spaces (e.g. the AusTrade Global Office Network). Crucially, proliferating globalising actor network often consist of intra- and inter-firm networks such as the well-connected KPMG-Centres of Excellence group.

The evidence from Australasian cities suggests the more pronounced emphasis on actors and places in Asia as prospective targets in international networking strategies. This spatial transition can be demonstrated by the 'Asianisation' of sister-city, or twinning, agreements; those cooperative agreements between places in geographically and politically distinct areas to promote cultural and commercial ties. A senior Auckland policy maker, for example, highlights the high importance of developing sister-city relationships with Asia's so-called Golden Triangle (China, Japan, Korea) for investment, tourism and higher education export purposes.[6] Notwithstanding the traditional objectives of enhancing international understanding and cultural awareness remaining important in sister city agreements, it appears that the newer relationships based on ties with Asian cities pursue more explicitly economic objectives of trade, investment as well as knowledge and technology sharing. Sister-city relationships, however, are increasingly questioned by local decision-makers as the created economic and cultural value in relation to the significant resources spent for overseas trips cannot be easily demonstrated, and therefore justified, to local constituents. It is important to note that the meta-governmental theme of 'Asianisation' is not just reflected in policy makers' and corporate elites' foci, but also visible in the complex spatial economic and cultural links of Asian immigrants such as ethnic food businesses in Sydney that source much production resources from their home countries.

These empirical findings on networking as important governmental practice confirm contributions from economic geography and related disciplines to the international literature that highlight the rise of network modes of resource allocations such as local growth alliances, multiple governance networks and self-organising governance. Associative practices such as social and spatial networking, project management, communication and dialogic forms of negotiation as well as learning and co-learning can be understood as vital ingredients in aligning governing actors and achieving (temporary) convergence of interests and actions in this regard. The research results also speak to recent academic work conducted on city-twinning. This governmental approach has been described as a commonplace practice that is central to strategies to enhance a city's international profile and global competiveness (Jayne, Hubbard and Bell 2011). It can be considered important to setting in motion a variety of global flows of people, ideas, money and goods that in the process form global urban networks. Importantly, the reason behind twinning is said to move from facilitating relationships based on political solidarity and humanitarianism to those which promise mutual economic value. Finally, it can be argued that global strategic network building in Australia and New Zealand is about building territorial and economic competiveness by creating productive and creative relationships that can facilitate economically viable interactions with global actors and places in general, and those located in increasingly powerful Asian countries, in particular. This tendency means that enhancing *global connectedness*, rather than, for example, knowledge-based

---

6    Research interview with the Auckland City Chief of Staff on 14 April 2010.

innovation, a new 'green deal' or broader sub-national resilience-building, is not just becoming the prevailing governmental rationality for New Zealand (Larner, Le Heron and Lewis 2007), but indeed for the whole Australasian region.

## Conclusions

This chapter illuminated key governmental strategies and practices that reconstitute 'glurbanising' territorial relations. From an Australasian urban perspective, it was argued that there are proliferating attempts to create a globally competitive investment environment for economic agents by means of new imaginaries, calculative practices, best practice and policy import, sustained urban place-making and marketing efforts, and strategic network building. Much emphasis in attempts to forge local–global connections is directed towards emerging markets in Asia; especially in China and India. A powerful governmental rationality is thus emerging in the form of *global connectedness* of actors, places and spaces that is seen as facilitative of the creation of globally competitive cities and city-regions. This logic builds on the now widely accepted assumption that in the context of ageing and thus talent-hungry western societies, the attraction of a skilled and exceedingly globally mobile workforce in order to produce goods and services for the global market place (or simply increase returns from local real estate investments), is a paramount territorial intervention goal. But the attraction of global economic resources also include events, conventions, tourists, students, migrants, headquarters, sister-city links, research and development facilities, foreign direct investment and so forth. In this regard, 'soft' spatial governmental strategies can be understood as establishing the 'worthiness' of a city to attract such scarce resources. The nature and the degree of local-global interactions under the 'glurbanisation' label is determined by local, path-dependent factors such as the economic base, place-specific cultures and local institutions, but follows also some general cross-regional trends such as the 'Asianisation' of new linkages and the emergence of new business-driven governance agents such as 'Committee for [City]' entities.

To return to the main book theme, one could argue that a 'governmental transition' is linked to a proposed 'industrial transition' in a sense that – viewed from the 'edge' of the world economy – new local-global spaces of economic management are constituted by 'soft', at-a-distance forms of influence that build on persuasion, association and calculation rather than on material incentives or top-down regulation. These initially discursive strategies attempt to reinforce local-global processes that are considered beneficial to fuel key capital accumulation strategies such as the expansion of highly profitable primary sector export activities. 'Glurbanising' cities are playing a central role in this governmental transition as they aim to reposition themselves favourably in the spatial division of production (Perth as global business services hub), consumption ('liveable' Melbourne), international finance ('Brand Sydney'), or are constructed as national exemplar of 'global visibility' in the case of an amalgamating Auckland.

To what degree these 'soft' globalising governmental strategies actually translate into spatially expanding economic processes, however, and the subsequent effects of possible changes in the investment environment on territorial outcome patterns, are largely unknown to date and should thus be subject to new empirical and context-sensitive research. While the notion of 'glurbanisation' is very useful in linking discursively framed urban-centred interventions with the new realities of globalising accumulation regimes in the Australasian context, there is a need for a more explicitly theorised and conceptualised discussion in the international literature on how spatial imaginaries are developed and deployed in globalising urban governance arenas.

Three swift comments deserve final attention. First, while the focus has been on the constitutive power of 'soft' interventions in cities, the nation-state still holds formidable regulatory power that must not be written off too easily. Substantial taxing powers in order to pay, amongst others, for nation-building infrastructure, education, health, welfare and defence, lie with the central/federal governments of the two scrutinised countries. In addition, discursive interventions such as the 'Brand New Zealand' initiative illuminate the ongoing importance of national institutions in contemporary governance practice. The second concern lies with the narrow cultural base for spatial references in Australasian cities. This study illuminates the fact that too often new ideas, models and benchmarks are confined to places in the Anglo-phone world. Potentially rewarding views for progressive territorial interventions in areas such as liveability, sustainability and vibrancy should reach beyond familiar territory and known knowledge and must simply travel the cultural-institutional distance needed to affect positive change. Third, the current governance and governmental shifts in the name of the 'global' seem to suggest that global economic connections are potentially privileged over local community and neighbourhood ties and processes. Auckland's lack of adequate local representation in its new 'Super-City' is just one example, the emergence of non-democratic planning institutions in Western Australia another one. Must globalization occur at the extent of democracy? This point raises the importance of framing globalising governance changes not as inevitable, but as complex and contested societal processes that always offer political choices. One of academic knowledge production's chief roles must be to make these political alternatives visible and enactable.

## Acknowledgements

The author acknowledges the valuable research contributions by Janine Hatch and Julie Cammell from the School of Earth and Environment at the University of Western Australia as well as the excellent technical assistance from Louise Holbrook. He is also grateful for valuable feedback on an early draft by Richard Le Heron from the University of Auckland. Special thanks to the very helpful comments of two anonymous reviewers.

## References

Boudreau, J. A. 2007. Making new political spaces: mobilizing spatial imaginaries, instrumentalizing spatial practices, and strategically using spatial tools. *Environment and Planning A,* 39(11), 2593–611.

Brenner, N. 2003. 'Glocalisation' as state spatial strategy: urban entrepreneurialism and the new politics of uneven development in western Europe, in *Remaking the Global Economy,* edited by J.Peck and H.W.C.Yeung. London: Sage, 197–215.

Brenner, N. 2004. *New state spaces: urban governance and the rescaling of statehood.* New York: Oxford University Press.

Bristow, G. 2005. Everyone's a 'winner': problematising the discourse of regional competitiveness. *Journal of Economic Geography,* 5, 285–304.

Committee for Perth 2010. *Research* [Online]. Available at http://www.committeeforperth.com.au/research [accessed: 11 October 2011]

Committee for Sydney 2009. *Global Sydney: Challenges and Opportunities for a Competitive Global City.* Benchmarking Paper, Committee for Sydney, Available at http://www.sydney.org.au/newsite/Benchmarking_Sydney_webdoc.pdf, [accessed: 11 October 2011]

EECW, 2009. *Conference 2009: Crisis, Opportunity and the New World Order.* Invitation/ Promotional Material for 'In-the-Zone' conference, EECW – Events, Meetings, Australia/ Perth (copy available from author).

Hodson, M. and Marvin, S. 2007. Understanding the Role of the National Exemplar in Constructing 'Strategic Glurbanization'. *International Journal of Urban and Regional Research,* 31(2), 303–25.

Jayne, M., Hubbard, P. and Bell, D. 2011. Worlding a city: Twinning and urban theory. *City,* 15(1), 25–41.

Jessop, B. 2002. Liberalism, neoliberalism, and urban governance: A state-theoretical perspective. *Antipode,* 34(3), 452–72.

Jessop, B. and Sum, N.L. 2000. An entrepreneurial city in action: Hong Kong's emerging strategies in and for (inter-) urban competition. *Urban Studies,* 37, 2287–313.

Kavaratzis, M. 2005. Place branding: A review of trends and conceptual models. *Marketing Review,* 5(4), 329–42.

Kennewell, C. and Shaw, B.J. 2008. Perth, Western Australia. *Cities,* 25, 243–55.

Larner, W. and Le Heron, R. 2005. Neo-liberalizing spaces and subjectivities: Reinventing New Zealand universities. *Organization,* 12(6), 843–62.

Larner, W., Le Heron, R. and Lewis, N. 2007. Co-constituting 'after neoliberalism': globalising governmentalities and political projects in Aotearoa/ New Zealand, in *Neoliberalization: States, Networks, People,* edited by K. England and K. Ward. London: Blackwell, 223–47.

Lee, J. 2010. Brand Sydney seeks Holmes a Court. *Sydney Morning Herald,* 27 March 2010, Available at http://www.smh.com.au/travel/travel-news/brand-sydney-seeks-holmes-a-court-20100326-r32f.html    [accessed:    11    October 2011]

Le Heron, R. and Pawson, E. 1996. *Changing Places. New Zealand in the Nineties.* Auckland: Longman Paul.

Matusitz, J. 2010. Glurbanization theory: an analysis of global cities. *International Review of Sociology,* 20(1), 1–14.

McCann, E. J. 2003. Framing Space and Time in the City: Urban Policy and the Politics of Spatial and Temporal Scale. *Journal of Urban Affairs,* 25, 159–78.

McCann, E.J. 2008. Expertise, truth, and urban policy mobilities: global circuits of knowledge in the development of Vancouver, Canada's 'four pillar' drug strategy. *Environment and Planning A,* 40, 885–904.

McGuirk, P. 2007. The political construction of the city-region: Notes from Sydney. *International Journal of Urban and Regional Research,* 31, 179-87.

O'Connor, K. 2003. Melbourne 2030: A response. *Urban Policy and Research,* 21(2), 211–15.

Sassen, S. 2009. Cities in Today's Global Age. *SAIS Review,* 29(1), 3-34.

Swyngedouw, E. 1997. Neither global nor local: 'glocalization' and the politics of scale, in *Spaces of Globalization: Reasserting the Power of the Local,* edited by K. Cox. New York: Guilford Press, 138–66.

Wetzstein, S. 2008a. Relaunching Regional Economic Development Policy and Planning for Auckland: Remaking the State and Contingent Governance under Neoliberalism. *Environment and Planning C,* 26, 1093–112.

Wetzstein, S. 2008b. Influencing Auckland's global economic participation? 'After-neoliberal' policy practices and regulatory effects on New Zealand's largest regional economy, in *Globalising Worlds and New Economic Configurations,* edited by C. Tamasy and M. Taylor. Aldershot: Ashgate, 271–85.

Wetzstein, S. 2010. *Perceptions of Urban Elites on Four Australasian Cities: How does Perth compare?* FACTBase Bulletin 18, The University of Western Australia and the Committee for Perth, Perth; Available at http://www.committeeforperth.com.au/research [accessed: 11 October 2011]

Wetzstein, S. and Le Heron, R. 2010. Regional Economic Policy 'in the Making': Imaginaries, Political Projects and Institutions for Auckland's Economic Transformation. *Environment and Planning A,* 42, 1902–24.

Chapter 14

# Conclusion: Towards a Refined Conceptualization of the Industrial Transition Approach in a Global-Local Context

Martina Fuchs and Martina Fromhold-Eisebith

## Introduction

Finally, we reach an overarching evaluation of what has been achieved by the set of chapters compiled in this book. Based on the empirical evidence reported by the contributors, we can identify major underpinnings that support the assumption of a newly emerging era of internationalized production that may be termed 'industrial transition'. It is now possible to highlight which processes appear to be most influential in shaping the present – and potentially also future – trajectories of industrial development from a local-global perspective. This final chapter will also revisit and refine conceptual assumptions associated with the 'industrial transition' framework, pointing out major merits of this approach as well as avenues for further research. Finally, we reflect briefly on political implications with respect to issues of government and governance too, in order to offer some ideas on what may form adequate responses to the new challenges at stake and ahead.

If we examine interpretations of the 'industrial transition' notion using advanced Internet search engines, we mainly find divergent and vague connotations of changes on the shop floor of factories, of new industrial landscapes, and some connections with the post-industrial society. Obviously, 'industrial transition' is not (yet) an established term bearing a meaning that is commonly understood. The contributors to this book have used 'industrial transition' as a new expression that puts into terms the ongoing changes resulting from the global-local interplay of various processes concerning manufacturing and services, economy and society, revealing their complexity. This raises the question if and to what extent the new 'industrial transition' concept is capable of and productive in advancing academic debates. The topics of advantages, deficits and further research horizons must therefore lie at the heart of this chapter.

We are convinced that the information assembled in this book delivers various good reasons for assuming the advent of a new era of 'industrial transition'. Principally, this era had its main origins in the 1980s and 1990s (with some early roots reaching decades or even centuries further back), but has gained enormous momentum during the last decade. The book chapters offer insights into diverse facets of relevant changes, such as the high amplitudes of economic cycles that are at the same time varying on the global level, the importance of knowledge and the influence of the financial sector, further the increasing flexibility that marks production, services and employment. These changes involve shifts in the power relations and the emergence of new institutional arrangements on the international, national and local scales.

At the same time, the term 'industrial transition' helps us to understand the changing links between various processes in different parts of the world. The concept highlights the role of new, strong industrial players in the world system (for example China, India and Brazil). These countries, on the one hand, threaten the traditional Western economies by increasing the global competition of prices and innovations, and, on the other hand, support the mature core regions by offering investment opportunities and new market areas, and by their overall economic dynamics.

There are, however, good reasons not to overstretch the notion of 'industrial transition' and to overestimate its significance because of the distinct and substantial fuzziness still associated with the notion, at least temporarily. Nevertheless we believe that a useful conceptual term has been coined that sets a spotlight on the quite special and unprecedented qualities of current global-local developments. This new terminology bears good potential to raise the awareness of economic geographers and academics of other scientific fields as well as in the media for the emergence of remarkably new socio-economic constellations and dynamics.

**Empirical Support for 'Industrial Transition' Qualities**

The book chapters have shed light on important drivers, process fields and constituents that form major backbones of the 'industrial transition'. And they have, above all, been able to illustrate the multi-faceted connections that link dynamics in different process fields, actor groups, and economic spaces. These aspects deserve some elaboration:

*Drivers and Major Constituents*

While the chapters in this volume provide evidence of the broad range of aspects and phenomena that earmark the 'industrial transition', they also reveal some recurring themes which indicate the particularly influential drivers and implications of this era of industrial change. Most notably, the economic crisis starting in 2008 and the induced flexibilization and evolution of organizations' (collaboration) strategies occupy centre stage, illustrated by a majority of authors (Chapters 3, 4, 5, 6, 8 and 9). While the crisis cannot actually be blamed for being the main initiator of 'industrial transition' trends, it has undeniably worked as a catalyst that magnifies, intensifies and accelerates processes that were instigated beforehand. Through the crisis, 'industrial transition' features have become particularly clear-cut and visible.

In line with that, the urge of companies and other organizations to gain flexibility and response capacities in the face of ever faster changing environments is another prominent topic. On the one hand, networking and collaboration activities emerge as a favoured strategic option, as shown by the linking up with local institutional assets by mechanical engineering firms (Fuchs and Kempermann, Chapter 5), or alliances between multinational subsidiaries and their local knowledge-intensive service partners (van Grunsven et al., Chapter 8). On the other hand, new configurations of the exploitation of assets have developed, as exemplified by the behaviour of financial investors towards automotive supplier firms (Scheuplein, Chapter 4), that of labour unions towards their clientele (Meyer, Chapter 6), or by intra-firm adjustments (Bathelt and Munro, Chapter 9).

Public policies and governance processes represent another set of drivers for 'industrial transition' highlighted in this book. A major issue is how regional policies in mature economies can be tuned towards new qualities of global competition challenges. Several ways are tried in order to cater to these challenges, inducing regional changes in industry structure and attitudes. Agency options include, among others, cluster initiatives (Hahn, Chapter 3), revised institutional support for certain knowledge sectors (Birch and Cumbers, Chapter 7), or drawing on 'soft' assets for urban marketing purposes (Wetzstein, Chapter 13). Altogether, the apparatus used for purposefully positioning a regional economy in the volatile 'industrial transition' environment seems to have broadened over recent years.

A last major driver of industrial change that receives particular attention in this book is, of course, the widening role and scope of industrial activities in newly emerging economies in Asia. Particular interesting in this regard: Contributions go beyond simply depicting international production shifts, but point out the socio-cultural underpinnings and consequences of these processes, be it in terms of the agency of transnational entrepreneurs (Henn, Chapter 10) or the gradual adoption of Western music styles and instruments by the Asian middle class (Chien et al., Chapter 11). The intricate interaction of organization changes, functional upgrading and relocation in one of the most dynamic economic spaces in China (Schiller, Chapter 12) is highly emblematic for what the 'industrial transition' means for regions that currently appear to be winners of global production shifts, but are also affected by the volatility of global conditions.

A commonality of virtually all of the processes and cases discussed in this book is their reference to multi-scale and multi-location patterns of interacting factors of change. We will turn to this in the following sections.

*Linking the Macro-Systemic, Micro and Meso Perspective*

The contributions illustrate that 'industrial transition' involves the systemic (or 'structural' or 'macro') level of the socio-economic sphere as well as the micro perspective on agents and actors. The concept also includes the meso view on regions with the focus on local politics, planning, administration and non-governmental organizations.

As a link between the systemic, individual and regional perspective, 'industrial transition' is a driving force for the ongoing changes and processes as well as a result of these dynamics. Because of these interdependencies, it is not possible to give a *general* overview of the reasons for the 'industrial transition' and its outcomes. Yet, in *specific cases* the causes and effects become obvious. For example, Hahn (Chapter 3) illustrates that there are strong effects of the macro level of worldwide competition on the micro level of the companies and their innovation activities, with – in that case – weak effects for regional interaction resp. inter-regional, cross-border linkages of firms.

Some authors, such as Fuchs and Kempermann (Chapter 5) and Meyer (Chapter 6) focus on employees as actors. They illustrate the strong effects of the macro level (the economic recession and institutional arrangements) on the micro level of the employees. Both chapters show, in contrast to Hahn (Chapter 3), that the meso level of the region can provide the relevant potential of flexibility for the companies and their workforce to overcome critical situations, including positive and negative effects for different segments of the labour market resp. for different groups of workers.

Last not least, actors on the micro level can be the relevant driving forces for changes on the macro and meso levels, as Henn (Chapter 10) shows by revealing how transnational entrepreneurs cause a global shift of production in diamond manufacturing, and how regional clusters participate in diamond fabrication and trade.

*Linking Different Economic Spheres and Sectors*

This book illustrates that 'industrial transition' also transcends the boundaries of changes in manufacturing alone. Scheuplein (Chapter 4), for instance, highlights the crucial impact of financial investors and their strategies on industrial restructuring. Anyway, it proves valuable to deepen the discussion about 'industrial transition' with regard to the financial sphere, as suggested by Taylor (Chapter 2). The power of financial corporations and institutions is increasing: new agents, such as rating agencies which are not legitimized by the citizens of a country, bear a growing influence on national and international policies and thus on politicians who were elected democratically. Furthermore, the topic of jurisdictional laxity (including tax havens) and the lack of international regulations of the financial markets remain virulent problems.

While we are editing this book, the crisis of the Euro dominates newspaper headlines and media reports. The problems of the Greek economy as well as other Southern European countries even raise the question about the future of the common European currency. Again, rating agencies play an important role in the serious game. At the same time, recent tendencies show the increasing importance of international political agreements and governmental harmonization on the international level, inducing some tension between national, supranational and transnational coordination and 'governance'.

Besides the aggravating effects of finance on the economy and society as a whole, the interdependencies between services and manufacturing are transformed. This becomes most obvious in the analysis provided by van Grunsven et al. (Chapter 8) on the strategic coupling between multinational subsidiaries and knowledge-intensive business services. In general, linking different economic spheres and sectors seems to be a productive research area within the 'industrial transition' framework. This refers to various topics, many of which could not be included in this volume, such as the role of shared knowledge and open innovation (Chesbrough 2003).

*Linking Risks and Resilience*

In line with deepening global-local development interdependencies, academic debates have put a strong focus on issues like global value chains, global production networks and global governance for many years (Coe, Dicken and Hess 2008). Yet, the specific topic of risks resulting from global-local relations has been widely marginalized in economic geography debates, leaving this terrain to business administration scientists. In large part, their studies advise company managers how to deal with risks, and there is a fluent passage to consulting business with topics like acquisition of information, contract design and insurance coverage (Manuj and Mentzer 2008: 133–37, Wakolbinger and Cruz 2011). The fact that many of our book contributions refer to crisis, recession and the like (as mentioned above), which evoke unexpected problems for regional actors and society, underscores the requirement to find academic answers to questions of risk in our globally interconnected world.

Natural disasters (e.g. the ash cloud of the Icelandic volcano Eyjafjallajoküll) are unpredictable local events that create sizeable economic problems on the local and international levels, causing negative effects in very different parts of the world. Correspondingly, the modern global society is highly dependent on complex and sensitive information technology and electronically controlled infrastructure. In 2011, the catastrophe in Japan has once again demonstrated that technology – nuclear power production – itself becomes an imminent source of danger and exacerbates unfolding calamities. The catastrophe has affected urban agglomerations with a high density of population, companies and infrastructure. Hence, it is the combination of our dependence on (often dangerous) technologies and the expansion of urbanized spaces that causes a high vulnerability with regard to unexpected disasters.

Apart from these risks posed to modern society and economy, some organizational changes have, since the 1980s and 1990s, made international production networks more vulnerable. Many firms have introduced lean production with a high degree of outsourcing and just-in-time delivery. They have increased global and single sourcing, often receiving specific parts only from a single supply firm. While the implementation of sourcing routines varies between companies and national modes of production (Freyssenet and Lung 2004), we generally see a tendency towards cost-cutting, but increasingly risky strategies (Hahn, Chapter 3, and Fuchs and Kempermann, Chapter 5).

Beyond the evidence provided in this book, two examples demonstrate that especially single sourcing can imply high costs in case of supply shortages, directly and indirectly. At the end of the 1990s, a fire in a supply firm of Toyota forced the automobile giant to shut down 18 plants. While the direct costs of that damage were about 195 million US-$, the additional indirect costs, resulting from about 70,000 cars which the company could not produce and sell during the period, mounted to about 325 million US-$. Another example is that of a fire in a Philips semiconductor plant in Albuquerque, which made the factory's largest client Ericsson suffer from about 400 million US-$ of lost revenues. Nokia, another client, had alternative supply firms and could avoid shortages (Wakolbinger and Cruz 2011: 4063–64). This illustrates that sourcing strategies – the linkages and exchange conditions between companies and in global production networks, respectively – play a strong role when it comes to risks and resilience.

Sometimes institutional arrangements are able to cushion the locational and regional effects of critical situations. These settings hardly influence the risks of global-local dependencies themselves, but do affect their impact. 'Industrial transition' therefore creates new arenas for actors on different spatial scales to adapt to, cope with, and shape the international risks (Christopherson, Mitchie and Tyler 2010, Simmie and Martin 2010). Analysing risks and resilience on different spatial scales seems to offer productive research opportunities for understanding the potential, assets and capabilities for coping with the 'industrial transition' and the involved dangers.

As several book chapters demonstrate, resilience is often bound to normative settings. Capacities for locational or regional resilience result from the influence of politics and interest groups that set an agenda for regional change, as Birch and Cumbers (Chapter 7) illustrate with respect to changing economic governance and institutional arrangements in the Scottish life sciences. This proves that, at least in some regions, 'industrial transition' opens small, but significant opportunities for policy-makers to position their region in global competition. Some regions have shaped particular forms of 'industrial transition', pursuing particular knowledge oriented policies in order to encourage and support certain types of industries, creating winners and losers. Wetzstein (Chapter 13) illustrates the case that especially urban areas try to improve their position in global competition by influencing international discourses, imaginaries and common patterns of interpretation. Thus, regional policy shapes the arena for local resilience.

*Linking Economy with Society and Ecological Environment*

The book contributions reveal that the 'industrial transition' is associated with a variety of economy related attributes: The continuous development of companies and sectors that operate in multi-location settings, the hybridisation of Fordist and post-Fordist features of production and the combination of 'high tech' and 'low tech' production, the prevailing trends towards increasing mechanization and electronification and, hence, towards sophisticated tasks (Chapters 5, 9, 11 and 12). At the same time, these trends interact with society and the ecological environment in multiple ways. On the one hand, 'industrial transition' requires resources from society and nature (exploiting raw materials and natural resources, using workforce, influencing policy for adjusting institutional settings etc.). On the other hand, 'industrial transition' is a product and result emerging from these conditions. First, the social and ecological (resp. natural) resources are limited. Political decisions and laws set restrictions to economic action. 'Industrial transition' performs inside of such boundaries and is shaped by such limitations. Second, 'industrial transition' is an answer to such constraints. New exploration technology for stretching the reserves, new fuel saving technologies for cars and diverse innovations to produce renewable energy, to transport, store and save energy are reactions to shortages as well as drivers of 'industrial transition'.

This unveils a fairly paradoxical, in some senses dialectic quality of this notion. The 'industrial transition' over time destroys assets that it has helped to create and make use of (resources, industrial activities). Yet, its actors also engage to some extent in 'healing' deficits, replacing overused and costly resources by cheaper, more abundant ones, and finding more rational ways for utilizing scarce production factors. It may be emphasized as a particular conceptual quality of this term that it makes us aware of such contradicting trends in economy and society.

## Terminological and Conceptual Refinement

We can hardly expect all economic geographers or scholars of other disciplines to understand and interpret the 'industrial transition' term just the way we do. Even the contributors to this book differ with respect to their interpretations of the notion due to their different backgrounds, nationalities and research perspectives. The various views taken on the conceptual framework, however, should not be conceived as jeopardizing its validity, but as constructive steps towards further conceptual refinement.

When looking through the chapters, we find inspiring variations of how the 'industrial transition' idea is associated with research objectives and outcomes. This includes some remarkable terminological amplification. Taylor (Chapter 2) uses the present participle 'industrial transitioning', which puts an even stronger emphasis on the transient and dynamic character of recent trends. Van Grunsven et al. (Chapter 8) scale our conceptual term down to the variant of 'regional transition', elucidating the embedding of localized processes in global economic frameworks.

Others distinguish between different categories of 'industrial transition', dividing, for instance, between constant incremental adjustments in production and unexpected, sudden or more drastic shifts (Bathelt and Munro, Chapter 9).

Other subdivisions are formed by separating 'spatial' from 'organisational' and 'institutional transitions' (Schiller, Chapter 12), which all refer to specific territorial manifestations of change created by actors and processes interacting across spatial scales. In the same context, even a 'double transition' of both outsourcing/offshoring strategies and a comprehensive overhaul of the value chain is stated. The nexus of 'spatial and organisational transitions' can be global-local, cross-border, or relating to governance modes. The variant of 'spatial transition', however, may also be interpreted in a broader perspective (which corresponds to our own conceptual understanding), as the emergence of new orientation marks ('Asianiation') in urban marketing policies (Wetzstein, Chapter 13). 'Governmental transition' is here seen as the trend towards 'soft' and distant forms of urban management.

We also observe some interesting oscillation between skeptical and optimistic connotations of the 'industrial transition' notion throughout this book. Some authors regard it as restricting the agency of policy-makers, but at the same time opening up options for new forms of economic governance (Birch and Cumbers, Chapter 7), for altered strategies of city regions to influence the investment environment (Wetzstein, Chapter 13), or for the agency of new types of transnational entrepreneurs (Henn, Chapter 10). 'Industrial transition' is surprisingly often connected with positive ideas of regional restructuring and upgrading through interactive knowledge, learning and innovation, integrated into global value chain dynamics (Chapter 7) and promoted by the activities of multinational companies (van Grunsven et al., Chapter 8). Societies also advance by virtue of 'industrial transition' with respect to changes in lifestyle and cultural orientation (Chien et al., Chapter 11). This stands in contrast to the gloomy record compiled on detrimental aspects of the 'industrial transition' by Taylor (Chapter 2).

Altogether these different perspectives and terminological variants underscore the multifaceted nature and contradictions that are inherent to the recent era of industrial change, hence the notion of 'industrial transition'. A task for the future will be to collect more empirical evidence in order to gradually complete the picture of countermoving forces and paradoxical constellations that proliferate in local-global development interdependencies. Further research should also pay more attention to topics that have only vaguely been covered in this book, such as systems of financialization, 'cooperative competition' in innovation, and multi-scalar constellations of risk and resilience. The overarching framework of the 'industrial transition' concept will be useful in providing the essential cognition that all of these processes are intricately connected.

## Policy Implications of the 'Industrial Transition'?

Obviously, 'industrial transition' is a field of engagement, clashes, struggles and fights, and the conflicts will continue. Furthermore, it is an arena where just a minority of participants seize an active part, while other are passively affected.

Only some of the latter will be able to gain a voice in the future. Up to now, many societies and people had to accept that industrial development implies the exclusion from wealth of large parts of the world population. It seems that industrial growth and those who mostly profit from it have largely ignored the moral right to appropriate living conditions for other creatures, be it humans or their 'relatives' in wildlife or animals used for 'food production'. As elaborated by Taylor (Chapter 2), there is something deeply unfair about power constellations in the 'industrial transition', raising the question of policy options.

With respect to the social and ecological interdependencies of 'industrial transition', we move towards areas of normative settings. What are the main directions of 'industrial transition' and where do these paths lead us? There will be different purposes and goals. Which ways are 'right' with regard to democratic participation, human rights and working conditions, and referring to ecological sustainability? Which countries, regions, companies and groups or individuals are the winners, and who are the losers?

There has been a long lasting critique against modernization theory with regard to the question if modernization is the 'right' path, and if all countries should follow it. The theoretical concept of 'industrial transition' is much more complex than this theory because it offers the insight into diverse ways of change and explicitly asks for theoretical and political presumptions. Still, we have to face up to the normative question about industrial transition: what direction and for whom?

The fact that the volatility and multiple qualities of the 'industrial transition' era go in line with corresponding trends in government and governance strategies renders the policy question difficult to solve. There may no longer be the option to resort to uniform 'regional concepts' of the 'Territorial Innovation Model' family (Moulaert and Sekia 2003) for virtually any country or region. A first imperative for political actors to be derived from the 'industrial transition' approach therefore is to clearly become aware of the increasingly complicated and inconsistent dynamics of socio-economic development, which demand tailor-made sets of measures that can further evolve over time. In terms of a second policy imperative, the growing imbalances and injustices induced by self-reinforcing developments that mark the 'industrial transition' deserve particular attention, connected to the ethical dimensions mentioned above. A third imperative is to explicitly take account of scale-bridging development interdependencies when designing policies, which demands a broadening of the horizon and scope of measures for supporting industry, labour qualifications and innovation (Fromhold-Eisebith 2007). Finally, however, it will be the courage and creativity of individual actors that count.

## References

Chesbrough, H.W. 2003. *Open Innovation*. Boston: Harvard Business School Press.

Christopherson, S., Michie, J. and Tyler, P. 2010. Regional resilience: Theoretical and empirical perspectives. *Cambridge Journal of Regions, Economy and Society*, 3, 3–10.

Coe, N.M., Dicken, P. and Hess, M. 2008. Introduction: global production networks – debates and challenges. *Journal of Economic Geography*, 8(3), 267– 69.

Freyssenet, M. and Lung, Y. 2004. Car firms' strategies and practices in Europe, in *European industrial restructuring in a global economy,* edited by M. Faust et al. Göttingen: SOFI, 85–103.

Fromhold-Eisebith, M. 2007. Bridging Scales in Innovation Policies: How to Link Regional, National and International Innovation Systems. *European Planning Studies*, 15(2), 217–33.

Manuj, I. and Mentzer, J.T. 2008. Global supply chain risk management. *Journal of Business Logistics*, 29(1), 133–55.

Moulaert, F. and Sekia F. 2003. Territorial innovation models: a critical survey. *Regional Studies*, 37, 289–302.

Simmie, J. and Martin, R. 2010. The economic resilience of regions: towards an evolutionary approach. *Cambridge Journal of Regions, Economy and Society*, 3, 27–43.

Wakolbinger, T. and Cruz, J.M. 2011. Supply chain disruption risk management through strategic information acquisition and sharing and risk-sharing contracts. *International Journal of Production Research,* 49 (13), 4063–84.

# Index